MTP International Review of Science

Biochemistry
Series One

Consultant Editors
H. L. Kornberg, F.R.S. and
D. C. Phillips, F.R.S.

Publisher's Note

The MTP International Review of Science is an important venture in scientific publishing, which is presented by Butterworths in association with MTP Medical and Technical Publishing Co. Ltd. and University Park Press, Baltimore. The basic concept of the Review is to provide regular authoritative reviews of entire disciplines. Chemistry was taken first as the problems of literature survey are probably more acute in this subject than in any other. Biochemistry and Physiology followed naturally. As a matter of policy, the authorship of the Review of Science is international and distinguished, the subject coverage is extensive, systematic and critical.

The Review has been conceived within a carefully organised editorial framework. The overall plan was drawn up and the volume editors appointed by seven consultant editors. In turn, each volume editor planned the coverage of his field and appointed authors to write on subjects which were within the area of their own research experience. No geographical restriction was imposed. Hence the 500 or so contributions to the Review of Science come from many countries of the world and provide an authoritative account of progress.

Biochemistry Series One (12 volumes) and Physiology Series One (8 volumes) are being published in the period 1973–1975. The 33 text volumes and 3 index volumes comprising Series One of Inorganic Chemistry, Physical Chemistry and Organic Chemistry were published in 1972–1973. In accordance with the stated policy of issuing regular reviews to keep the series up to date, volumes of Chemistry Series Two are being published in the period 1975–1976. In Biochemistry Series One, Physiology Series One and Chemistry Series Two a subject index is incorporated in each volume and there is no separate index volume.

Butterworth & Co. (Publishers) Ltd.

ORGANIC CHEMISTRY SERIES TWO

Consultant Editor
D. H. Hey, F.R.S., *formerly of the Department of Chemistry, King's College, University of London*

Volume titles and Editors

1 STRUCTURE DETERMINATION IN ORGANIC CHEMISTRY
Professor L. M. Jackman, *Pennsylvania State University*

2 ALIPHATIC COMPOUNDS
Professor N. B. Chapman, *Hull University*

3 AROMATIC COMPOUNDS
Professor H. Zollinger, *Eidgenossische Technische Hochschule, Zurich*

4 HETEROCYCLIC COMPOUNDS
Dr. K. Schofield, *University of Exeter*

5 ALICYCLIC COMPOUNDS
Professor D. Ginsburg, *Technion-Israel Institute of Technology, Haifa*

6 AMINO ACIDS, PEPTIDES AND RELATED COMPOUNDS
Professor N. H. Rydon, *University of Exeter*

7 CARBOHYDRATES
Professor G. O. Aspinall, *York University, Ontario*

8 STEROIDS
Dr. W. F. Johns, *G. D. Searle & Co., Chicago*

9 ALKALOIDS
Professor K. Wiesner, F.R.S., *University of New Brunswick*

10 FREE RADICAL REACTIONS
Professor W. A. Waters, F.R.S., *formerly of the University of Oxford*

INORGANIC CHEMISTRY SERIES TWO

Consultant Editor
H. J. Emeléus, C.B.E., F.R.S.
Department of Chemistry University of Cambridge

Volume titles and Editors

1 MAIN GROUP ELEMENTS—HYDROGEN AND GROUPS I–III
Professor M. F. Lappert, *University of Sussex*

2 MAIN GROUP ELEMENTS—GROUPS IV AND V
Dr. D. B. Sowerby, *University of Nottingham*

3 MAIN GROUP ELEMENTS—GROUPS VI AND VII
Professor V. Gutmann, *Technical University of Vienna*

4 ORGANOMETALLIC DERIVATIVES OF THE MAIN GROUP ELEMENTS
Professor B. J. Aylett, *Westfield College, University of London*

5 TRANSITION METALS— PART 1
Professor D. W. A. Sharp, *University of Glasgow*

6 TRANSITION METALS— PART 2
Dr. M. J. Mays, *University of Cambridge*

7 LANTHANIDES AND ACTINIDES
Professor K. W. Bagnall, *University of Manchester*

8 RADIOCHEMISTRY
Dr. A. G. Maddock, *University of Cambridge*

9 REACTION MECHANISMS IN INORGANIC CHEMISTRY
Professor M. L. Tobe, *University College, University of London*

10 SOLID STATE CHEMISTRY
Dr. L. E. J. Roberts, *Atomic Energy Research Establishment, Harwell*

PHYSICAL CHEMISTRY SERIES TWO

Consultant Editor
A. D. Buckingham, F.R.S., *Department of Chemistry University of Cambridge*

Volume titles and Editors

1 THEORETICAL CHEMISTRY
Professor A. D. Buckingham, F.R.S.,*University of Cambridge* and Professor C. A. Coulson, F.R.S., *University of Oxford*

2 MOLECULAR STRUCTURE AND PROPERTIES
Professor A. D. Buckingham, F.R.S.,*University of Cambridge*

3 SPECTROSCOPY
Dr. D. A. Ramsay, F.R.S.C., *National Research Council of Canada*

4 MAGNETIC RESONANCE
Professor C. A. McDowell, F.R.S.C., *University of British Columbia*

5 MASS SPECTROMETRY
Professor A. Maccoll, *University College, University of London*

6 ELECTROCHEMISTRY
Professor J. O'M Bockris, *The Flinders University of S. Australia*

7 SURFACE CHEMISTRY AND COLLOIDS
Professor M. Kerker, *Clarkson College of Technology, New York*

8 MACROMOLECULAR SCIENCE
Professor C. E. H. Bawn, C.B.E., F.R.S., *formerly of the University of Liverpool*

9 CHEMICAL KINETICS
Professor D. R. Herschbach *Harvard University*

10 THERMOCHEMISTRY AND THERMO-DYNAMICS
Dr. H. A. Skinner, *University of Manchester*

11 CHEMICAL CRYSTALLOGRAPHY
Professor J. Monteath Robertson, C.B.E., F.R.S., *formerly of the University of Glasgow*

12 ANALYTICAL CHEMISTRY —PART 1
Professor T. S. West, *Imperial College, University of London*

13 ANALYTICAL CHEMISTRY —PART 2
Professor T. S. West, *Imperial College, University of London*

BIOCHEMISTRY
SERIES ONE

Consultant Editors
H. L. Kornberg, F.R.S.
*Department of Biochemistry
University of Leicester* and
D. C. Phillips, F.R.S., *Department of
Zoology, University of Oxford*

Volume titles and Editors

**1 CHEMISTRY OF MACRO-
MOLECULES**
Professor H. Gutfreund, *University of
Bristol*

**2 BIOCHEMISTRY OF CELL WALLS
AND MEMBRANES**
Dr. C. F. Fox, *University of California,
Los Angeles*

**3 ENERGY TRANSDUCING
MECHANISMS**
Professor E. Racker, *Cornell University,
New York*

4 BIOCHEMISTRY OF LIPIDS
Professor T. W. Goodwin, F.R.S.,
University of Liverpool

**5 BIOCHEMISTRY OF CARBO-
HYDRATES**
Professor W. J. Whelan, *University
of Miami*

**6 BIOCHEMISTRY OF NUCLEIC
ACIDS**
Professor K. Burton, F.R.S., *University of
Newcastle upon Tyne*

**7 SYNTHESIS OF AMINO ACIDS
AND PROTEINS**
Professor H. R. V. Arnstein, *King's
College, University of London*

8 BIOCHEMISTRY OF HORMONES
Professor H. V. Rickenberg, *National
Jewish Hospital & Research Center,
Colorado*

**9 BIOCHEMISTRY OF CELL DIFFER-
ENTIATION**
Professor J. Paul, *The Beatson Institute
for Cancer Research, Glasgow*

10 DEFENCE AND RECOGNITION
Professor R. R. Porter, F.R.S., *University
of Oxford*

11 PLANT BIOCHEMISTRY
Professor D. H. Northcote, F.R.S.,
University of Cambridge

**12 PHYSIOLOGICAL AND PHARMACO-
LOGICAL BIOCHEMISTRY**
Dr. H. K. F. Blaschko, F.R.S., *University
of Oxford*

PHYSIOLOGY
SERIES ONE

Consultant Editors
A. C. Guyton,
*Department of Physiology and
Biophysics, University of Mississippi
Medical Center* and
D. F. Horrobin,
*Department of Physiology, University
of Newcastle upon Tyne*

Volume titles and Editors

1 CARDIOVASCULAR PHYSIOLOGY
Professor A. C. Guyton and Dr. C. E. Jones,
University of Mississippi Medical Center

2 RESPIRATORY PHYSIOLOGY
Professor J. G. Widdicombe, *St. George's
Hospital, London*

3 NEUROPHYSIOLOGY
Professor C. C. Hunt, *Washington
University School of Medicine, St. Louis*

4 GASTROINTESTINAL PHYSIOLOGY
Professor E. D. Jacobson and Dr. L. L.
Shanbour, *University of Texas Medical
School*

5 ENDOCRINE PHYSIOLOGY
Professor S. M. McCann, *University of
Texas*

**6 KIDNEY AND URINARY TRACT
PHYSIOLOGY**
Professor K. Thurau, *University of Munich*

7 ENVIRONMENTAL PHYSIOLOGY
Professor D. Robertshaw, *University
of Nairobi*

8 REPRODUCTIVE PHYSIOLOGY
Professor R. O. Greep, *Harvard Medical
School*

MTP International Review of Science

Biochemistry
Series One

Volume 3
Energy Transducing Mechanisms

Edited by **E. Racker**
Cornell University

Butterworths · London
University Park Press · Baltimore

THE BUTTERWORTH GROUP

ENGLAND
Butterworth & Co (Publishers) Ltd
London: 88 Kingsway, WC2B 6AB

AUSTRALIA
Butterworths Pty Ltd
Sydney: 586 Pacific Highway, 2067
Melbourne: 343 Little Collins Street, 3000
Brisbane: Commonwealth Bank Building, King George Square, 4000

NEW ZEALAND
Butterworths of New Zealand Ltd
Wellington: 26–28 Waring Taylor Street, 1

SOUTH AFRICA
Butterworth & Co (South Africa) (Pty) Ltd
Durban: 152–154 Gale Street

ISBN 0 408 70497 7

UNIVERSITY PARK PRESS

U.S.A. and CANADA
University Park Press
Chamber of Commerce Building
Baltimore, Maryland, 21202

Library of Congress Cataloging in Publication Data
Main entry under title:
Energy transducing mechanisms.

(Biochemistry, series one; v. 3) (MTP
international review of science)
Includes index.
1. Energy metabolism. 2. Biological
transport. 3. Membranes (Biology) I. Racker,
Efraim, 1913–   II. Series. III. Series:
MTP international review of science. [DNLM:
1. Biophysics. 2. Energy transfer. W1 B1633
ser. 1 v. 3/QT34 E54]
QP501.B527   vol. 3   [QP171]   574.1′92′08s
[574.1′9283]
ISBN 0–8391–1042–1   75–8804

First Published 1975 and © 1975

Typeset, printed and bound in Great Britain by
REDWOOD BURN LIMITED
Trowbridge & Esher

Consultant Editors' Note

The MTP International Review of Science is designed to provide a comprehensive, critical and continuing survey of progress in research. Nowhere is such a survey needed as urgently as in those areas of knowledge that deal with the molecular aspects of biology. Both the volume of new information, and the pace at which it accrues, threaten to overwhelm the reader: it is becoming increasingly difficult for a practitioner of one branch of biochemistry to understand even the language used by specialists in another.

The present series of 12 volumes is intended to counteract this situation. It has been the aim of each Editor and the contributors to each volume not only to provide authoritative and up-to-date reviews but carefully to place these reviews into the context of existing knowledge, so that their significance to the overall advances in biochemical understanding can be understood also by advanced students and by non-specialist biochemists. It is particularly hoped that this series will benefit those colleagues to whom the whole range of scientific journals is not readily available. Inevitably, some of the information in these articles will already be out of date by the time these volumes appear: it is for that reason that further or revised volumes will be published as and when this is felt to be appropriate.

In order to give some kind of coherence to this series, we have viewed the description of biological processes in molecular terms as a progression from the properties of macromolecular cell components, through the functional interrelations of those components, to the manner in which cells, tissues and organisms respond biochemically to external changes. Although it is clear that many important topics have been ignored in a collection of articles chosen in this manner, we hope that the authority and distinction of the contributions will compensate for our shortcomings of thematic selection. We certainly welcome criticisms, and solicit suggestions for future reviews, from interested readers.

It is our pleasure to thank all who have collaborated to make this venture possible—the volume editors, the chapter authors, and the publishers.

Leicester   H. L. Kornberg

Oxford   D. C. Phillips

Preface

The historians of energy transducing mechanisms will perhaps refer to the years following 1970 as the period of conversion. Oxidative phosphorylation, ion transport, muscular contraction and membraneous excitability have been more or less duplicated in simple model systems and basic mechanisms of operation have been formulated. Although many of the details are still obscure we begin to recognise not only the moving parts but some remarkable similarity between them. Today it is no longer surprising to discover that the energy-generating machinery of chloroplasts, mitochondria and bacteria operates according to the same principle, or that proteins involved in the translocation of Na^+ and Ca^{2+} ions across membranes have analogous phosphorylated intermediates. But now similarities are emerging between energy transducing mechanisms as widely apart in function as the phosphorylating apparatus of mitochondria and the contracting fibril of muscles. Thus nature keeps imitating itself, providing us with ingenious variations of the same theme to fit the purpose of specialised functions. We see a new unity of biochemistry that extends not only horizontally through the various kingdoms of life, but vertically through the diversity of functions.

The chapters in this book describe areas in the frontiers of research in energy transducing mechanisms. The first chapter deals with methodological approaches to membrane function. Intrinsic (or natural) probes that undergo changes in spectral or fluorescence properties during function have been widely used in the past. Now extrinsic probes without toxicity are being made to order to fit the specific problem at hand. Electron spin and nuclear magnetic resonance as well as x-ray diffraction techniques are becoming increasingly useful. In the second chapter, energy transducing mechanisms that are responsible for the generation of energy during oxidative processes are discussed in depth. The chapter includes exciting new advances in the reconstitution of model systems that permit measurements of electrical events which until now have eluded the experimental approach. The third chapter deals authoritatively with muscular contraction with particular emphasis on the mechanism of action of myosin and actomyosin. Included are relations to active ion transport and control of muscular contraction. The fourth chapter is an exciting adventure into excitable membranes. Model systems are described which to a remarkable degree imitate the electrical events that occur during excitation of natural membranes. A final and brief chapter deals with control mechanisms of energy transducing systems, a sorely neglected area of research.

The emergence of common tools, such as membrane probes, in the experimental approach to different energy transducing systems has helped the conversion of widely separated areas of research. On coming closer to each other the investigators are beginning to discover not only a mutual interest in experimental methods but a remarkable unity of biochemistry across functional diversity.

New York   E. Racker

Contents

1
Electron Transport and Energy-dependent Responses of Deep and Shallow Probes of Biological Membranes

B. CHANCE
University of Pennsylvania

Abbreviations

ANS	1-anilinonaphthalene-8-sulphonate
AS	12-(9-anthroyl)stearate
diO-C_6(3)	3,3'-dihexyloxacarbocyanine
DPL	dipalmitoyl-lecithin
e.s.r.	electron spin resonance
MC-I	merocyanine I
MC-II	merocyanine II
MC-V	merocyanine III
MNS	2-(N-methylanilino)naphthalene-6-sulphonate
n.m.r.	nuclear magnetic resonance (also used for proton resonance studies)
Rps.	Rhodopseudomonas
T_M	melting point (temperature of melting)

1.1 INTRODUCTION

Increasing interest in and understanding of the static and dynamic aspects of membrane structure and function as studied by x-ray techniques[1] on the one hand and by fluorescent[2] or spin label[3] probes on the other can now be extended to the location and the nature of energy coupling reactions in biological membranes. Using as a working hypothesis the concept that electron transport and energy coupling cause massive structural and electrical effects in mitochondria, submitochondrial particles, chromatophores and chloroplasts, there must follow significant changes in membrane structure which can be detected by appropriate probes. These probes can be those such as carotenoid already present in the membrane, termed 'intrinsic' probes, or those added, 'extrinsic' probes which interact with the membrane to a sufficient extent to report what is happening there, and yet at the same time alter the membrane structure in only an innocuous way so that function is retained.

The type of structural change involved in energy coupling is unknown in detail, but a range of structural changes can be caused to occur in membranes by their responses to thermal transitions[4], to ion gradients[5], to swelling and shrinking[6] and a variety of other transients. These changes occur on large and small scales. On a large scale, they are observable in the general category of 'orthodox-condensed'[7] changes studied in a number of laboratories, particularly those of Hackenbrock[8] and Green[9]. (For a summary, see Ref. 10). Such changes may involve gross reorganisations of macroscopic units of the membrane and give little information on the underlying changes at the molecular level. On the other hand, membrane structure changes may be on such a small scale as to be undetectable by anything larger than a probe of molecular dimensions occupying the membrane either at the aqueous interface or in the hydrocarbon layer. It is on these small-scale changes that the interest of this contribution is focused, describing in general a number of the properties of these probes, the techniques for their readout, and the systems to which they may be applied, and concluding with some specific examples.

1.2 TYPES OF PROBE RESPONSE

A probe that does not interact with the biological system cannot report about the system. There may be two kinds of interaction; in the first, the occupancy of the membrane by the probe remains constant throughout the transitions involved, and in the second the occupancy changes with changing membrane conditions.

With constant occupancy, the probe reports motions of the environment with respect to the probe or of the probe with respect to the environment. An example of this phenomenon is provided by 12-(9-anthroyl)stearic acid (AS)[11] which occupies the membrane to the extent of 20 nmol per mg of protein (for submitochondrial particles) with a dissociation constant[12] of less than 10^{-6} mol l^{-1}. Thus, AS fulfills the criteria for a completely bound probe. AS is a 'deep' probe, occupying the hydrocarbon region of natural and artificial membranes; in both types of membranes, its fluorescence is quenched by oxidised ubiquinone[13]. Since the degree of probe quenching is proportional to the amount of oxidised ubiquinone, AS is an indicator of the redox state of this respiratory carrier and of its functionality and mobility in the membrane, i.e. AS responds to electron transport. It does not, apparently, respond to energy coupling.

8-Anilinonaphthalene-1-sulphonate (ANS)[14] is representative of the second type of probe, where the different degrees of membrane occupancy in the energised and de-energised states depend upon membrane structure, charge gradient, or membrane potential. It has been proposed that ANS, a probe of the aqueous interface[15], might enter the hydrocarbon phase to a greater extent in the energisation process. Similar responses of the diO-C_6(3) probe have more recently been reported in detail by Hoffman et al.[16]. In the case of differing extents of membrane occupancy by the probe, the information on membrane structure may be limited since in one state the probe may not occupy the membrane at all and thus differential probe responses to energisation and de-energisation or to electron transport cannot be obtained. In fact, some have termed such probes* when dissociated from the membrane, 'non-probes' or 'slow probes'.

Although ANS is useful as a probe of the energised state of the membrane and detects, with great sensitivity, energisation by either electron transport or ATP activation[17], it suffers from the fact that it responds both to the surface charge of the membrane[18] and to membrane structure changes. For example, surface charges cause a considerable increase in the binding constant of ANS in the energised state with respect to the de-energised state[18]. However, the extrapolated quantum yield also increases in the energised state; when the fluorescence intensity is extrapolated to the 'completely bound' state (i.e. a large membrane-to-probe ratio), the plot shows that the fluorescence intensity for the energised state is about twice as great as that for the de-energised state[19]. This phenomenon is attributed to a structural change in the membrane in the energised state which results in smaller quenching of the probe fluorescence. The two effects are difficult to distinguish, and under the usual experimental conditions the increased binding of the probe is the main cause of the increased fluorescence on membrane energisation. Radda and Vanderkooi[2] have discussed the parameters of ANS fluorescence responses in detail.

The ideal probe would be bound to the membrane to the same extent in the energised and de-energised states, and report charge and structure changes on energisation in the region of the membrane that it occupies. In pigeon heart mitochondria, ethidium bromide approaches this condition more closely that does ANS (Vignais et al., unpublished observations) although rat liver mitochondria give different results[20]. Merocyanine V appears to be tightly bound to the membrane in both states and shows large fluorescence and absorption changes characterising the energised state owing to changes of membrane charge[22]. The rationale for employing probes specifically designed to detect charge changes is afforded in the work of Platt[23, 24], Brooker et al.[25], Bücher et al.[26], and Schmidt and Reich[27].

* P. Mueller, personal communication.

The methods for observing the properties of the probes and, from them, evaluating the probe environment in both the static and dynamic sense include a wide variety of biophysical techniques, of which spectroscopy is pivotal. In fact, spectroscopy of different probes is possible at almost any wavelength of the spectrum and over a time range running from picoseconds to days or weeks.

1.3 TIME DOMAINS

Each spectroscopic technique has a particular time range in which it serves optimally, and one classification of these techniques is therefore based on these different domains.

1.3.1 Picosecond time range

Considering first the shortest time range, optical spectroscopy is useful down to a few picoseconds, as evidenced by recent studies of carotenoid absorbancy changes upon flash illumination of the reaction centre of *Rps. spheroides*[28]. In that case, the optical readout of the absorbancy change of the intrinsic probe can be made with a resolution of 10 ps. Changes that occur in this time range can be read out in real time by transmission spectroscopy, but not by fluorometry which may be significantly delayed. Thus, transmission spectroscopy far exceeds the time resolution of other techniques described below.

1.3.2 Nanosecond time range

Here both the electron spin resonance and fluorescence emission methods come into their own in their ability to report motional information. Fluorescence emission can report events from fractions of a nanosecond to hundreds of nanoseconds, as shown by the properties of the merocyanine dyes on the one hand[21, 29] and of pyrene butyric acid[30] on the other. Using e.s.r. techniques, the nitroxide labels are capable[31] of recording motional events from 10^{-9} s onwards to 10^{-7} s, with the possibility of considerable extension towards 10^{-3} s by measurements of the first and second harmonics of the modulation frequency[32].

1.3.3 Microsecond time range

The e.s.r. technique is not currently practical in the microsecond time range and fluorescent probes are usually beyond the upper limit of their range for reporting motional events. However, recent developments in the noise modulation of signals have been described by Magde *et al.*[33] and applied to artificial membranes; in this particular case, fluorescence measurements may give a better signal-to-noise ratio. Photo-induced dichroism (*cf.* Section 1.4.1.2) has proved very useful in identifying the motion of photo-active biological chromophores such as rhodopsin[34], chlorophyll[35] and, in preliminary studies, cytochrome oxidase rendered photosensitive by combination with carbon monoxide[36]. In this time range, flash photolysis of hemoglobin: CO and cytochrome a_3: CO in the presence of oxygen[37] affords a convenient method for starting the reaction, and the flash can usually be made non-interfering when used with absorption and fluorescence spectroscopy by an appropriate choice of wavelengths and optical filters.

1.3.4   Millisecond time range

Although fluorescence and absorption spectroscopy continue to be useful in the millisecond time range, nuclear magnetic resonance (n.m.r.) comes into its own here and provides an approach to the magnetic resonance spectra of hydrogen, phosphorus, carbon, fluoride and many other atoms of biological interest. In a similar time range, the latest developments of magnetic susceptibility apparatus based on superconducting detectors may approach the millisecond region[38], but are usually considered to be in longer time ranges even in the configuration of the Iskanderian balance employed by Brill[39] in his pioneering experiments on the magnetic susceptibility of catalase.

1.3.5   Longer time ranges

In the last time range is the technique of x-ray diffraction, which in its usual form may require hours or days for data collection; however, in a currently developing mode as used in the time-resolved study of muscle contraction[40] or of membrane structure changes[41, 42], position-sensitive detectors[43] used in a 'gated' configuration with repetitive techniques may collect data in much shorter times.

In summary, optical spectroscopy is the only technique which can cover the time range from purely physical to biochemical processes. As this contribution will develop, it is a technique for which a relatively firm theoretical ground exists for the study of electrochromic probes responsive to membrane potentials. Fluorescence signals, although often easier to obtain, are restricted in both time range and interpretation. E.s.r. signals, extremely useful for mobility studies in membranes, give results similar to fluorescence signals. N.m.r. signals, which have a relatively narrow time range, have proved highly significant for a number of membrane structure studies. X-Rays, of course, remain the standard for structural determination at the atomic level; however, at the present time the lack of homogeneity and repetitiveness of membrane structure, together with the natural fluidity of membranes, limits the resolution attainable with this approach; new possibilities for the time-resolved, gated mode of operation are only recently available.

Considering principally optical and fluorescence spectroscopy, the wavelengths of maximal absorption, excitation and emission spectra can be sensitive to the solvatochromicity of the probe, to the effects of solvents and of changes of dielectric constant, 'Z-value' and other parameters upon the spectroscopic characteristics of the probe. Probes whose fluorescence characteristics are of primary importance, such as AS and ANS, are examined for changes in fluorescence lifetime, quantum yield and polarisation which indicate changes of membrane structure. Electrochromic probes, such as the intrinsic carotenoids and extrinsic merocyanines, respond to electric fields according to linear, quadratic and more complex relationships.

1.4 PROBE PARAMETERS

1.4.1 Dynamic effects

1.4.1.1 Fluorescence characteristics

Dynamic effects related to the fluidity of the probe occupancy site and to the mobility of the probe itself are best identified from the fluorescence characteristics of probes such as ANS and AS. These include the quantum yield and the lifetime of the fluorescence emission, which are diminished by collisional processes based upon the diffusional properties of the molecules involved. On a steady state basis, the degree of polarisation of the light emitted by the probes[44] gives information related to their partial or complete rotation during the interval of the fluorescence lifetime.

1.4.1.2 Photo-induced dichroism

Photo-induced dichroism affords another approach of a much broader time scale and thus more general applicability in cases where the probe is also a 'photolyte', *i.e.* capable of an alteration in its properties in response to a light flash. Motion of the probe occurring between the light flash and the earliest measurement alters the polarisation of the light absorption. This change is recorded as a time-dependent increase of the population of the optical field when the polarisations of the photolysis and measuring light beams are parallel, or as a time-dependent decrease of detection of the photolyte where these polarisations are perpendicular to one another, provided the photolysis and absorption measurements utilise the same absorption band or when photolysis results in complete bleaching.

1.4.2 Distance measurements

Structure determinations by distance measurement between probes are finding greater applicability, and several approaches are available.

1.4.2.1 Proton magnetic resonance

Broadening of the proton magnetic resonance signals of membrane components by the close approach of a paramagnetic ion[45] is most useful in membranes where the ion can be localised, e.g. in the aqueous interface by binding to the phosphate head-group as is the case[46] with Mn^{2+}, or more generally by membrane impermeability to an ion such as gadolinium[47].

1.4.2.2   Electron paramagnetic resonance

Similarly, paramagnetic ions can broaden the e.s.r. signals of paramagnetic membrane components such as iron–sulphur proteins and oxidised cytochromes, provided they are accessible one to the other[48]. Mn^{2+} or Ni^{2+} would be expected to be especially useful for determining the location of the cytochromes which exhibit e.p.r. transitions in the $g = 3$ region, away from the Mn^{2+} or Ni^{2+} signal at $g = 2$; conversely, Gd^{3+}, with a broad e.p.r. signal at higher g values, is useful for studies of the iron–sulphur proteins of membranes derived from mitochondria[49] or *Chromatium*[50].

1.4.2.3   Resonance energy transfer

In the range from 20 to 80 Å for protein molecules and over shorter distances for smaller molecules where there is no overlap of molecular orbitals, resonance energy transfer between two probes, either extrinsic or intrinsic, affords a distance measurement when the energy transfer from one probe to another is verified both by a decrease in the fluorescence emission of one and an increase in the fluorescence emission of the other, and by a shortening of the donor lifetime. Since this range spans the roughly 50 Å width of the bilayer, this method may not be effective when components are to be located perpendicular to the plane of the membrane; however, their location in the plane of the membrane can be determined in a qualitative sense without a knowledge of the orientation of the molecules. Where this orientation is known, for example by determinations of the transition moment as described below, as in the case of AS[51] or ANS[12], this method can be quite precise. It is less precise but nevertheless useful where the orientation factor is unknown or must be assumed; on this qualitative basis, the energy transfer method indicates that merocyanine occupies 'patches' in the membrane which are within 50 Å of the sites occupied by ANS and AS[21].

1.4.3   Orientation of the probe

The orientation of the probe with respect to the plane of the lipid can be determined from optical measurements of the direction of the transition moment of the molecule in oriented bilayers. In the case of ANS, this is parallel to the plane of the membrane; in AS, one transition moment (350 nm) is parallel while the other (250 nm) is perpendicular[12]. These results, coupled with x-ray and n.m.r. data, align the two probes with respect to the plane of the membrane. The determination of the transition moment of the merocyanine probe by spectroscopy and fluorimetry indicates that the long axis of the molecule is approximately 30° from the plane of the lipid bilayer.

1.4.4 Sidedness of membranes

Several methods have been found useful in determining the sidedness of membranes, *i.e.* on which side of the membrane particular proteins and lipids are oriented[2]. For example, selective broadening of the e.p.r. signals of membrane components by paramagnetic ions to which the membrane is impermeable, such as Ni^{2+}, indicate that the iron–sulphur proteins are near the surface of the membrane.

1.4.4.1 Proton magnetic resonance

Similarly, the paramagnetic species ferricyanide will cause a shift in the resonant frequency of the lipid protons that can readily be detected by high-frequency proton magnetic resonance techniques; it is therefore termed a 'shift probe'. Since ferricyanide does not penetrate the lipid vesicles, it can

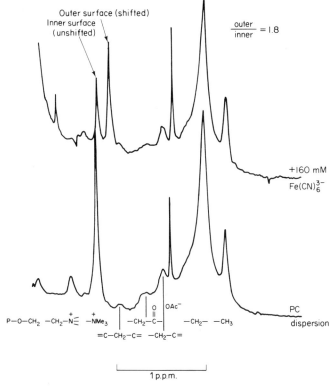

Figure 1.1 An assay of the outside/inside ratio of lecithin protons in a small vesicle fraction obtained from a Sepharose 4B column. Lower curve, egg lecithin (130 mM) in 20 mM sodium acetate, pH = 8.4. Upper curve, the effect of the addition of 160 mM ferricyanide upon the proton magnetic resonances at 270 MHz. (From Barker *et al.*[52], by courtesy of Elsevier.)

be used to determine the distribution of two types of phospholipids on either side of the membrane, as has recently been pointed out by Barker et al.[52]. In the example discussed by them, lipid vesicles formed from lecithin dispersions were fractionated into small vesicles of uniform size and treated with ferri-cyanide (Figure 1.1). Whereas the same concentration of dispersed lipid gives only a single peak, the diagram of Figure 1.1 shows 'outer surface shifted' and 'inner surface unshifted' proton resonances in the ferricyanide-treated vesicles with a ratio of the areas under the peaks of 1.8, which represents the ratio of the number of lecithin protons on the outside and on the inside of the vesicles. This ratio, being a function of the diameter of the vesicles, de-creases to 1.0 as larger vesicles are studied[52], owing to a redistribution of the lipids. The impact of this observation on the composition of the proteins in 'sonicated particles' is obvious; the segregation of the electron carriers in these vesicles may well be an artifact of asymmetric lipid distribution characteristic of the range of diameters of the submitochondrial particles obtained by sonication.

1.4.5 Membrane functionality

A last and precautionary word about probe properties concerns their relation-ship to membrane functionality. Usually, optical methods are sufficiently sensitive that probe occupancy levels can be kept below those that affect any function significantly. However, delicately organised functions such as energy coupling and ion transport are more sensitive to the presence of the probe; these functions of the system should be compared in the presence and absence of the probe. In cases where primary events such as electric field generation may be involved, physical effects are predominant and chemical modification of the system might not be expected to alter anything more than the distances between the probe and the system to be studied. In x-ray studies, however, the probe occupancy must be increased to the point where the lipid structure is distorted, since the probe per se is not observable by x-rays; here, suitable controls are required to ensure careful monitoring of the functional changes caused by these detectable alterations of membrane structure.

1.5 PROBE LOCATION BY FLUORESCENCE EMISSION SPECTROSCOPY

1.5.1 Response to the dielectric constant of the medium

To a first approximation, the location of a fluorescent probe in natural and artificial membranes can be determined by its response to the dielectric constant of the medium it occupies. Figure 1.2 compares data on the fluores-cence emission of AS and ANS in biological membranes. Whereas these probes have similar characteristics in solvents of high dielectric constant, their response differs considerably in membrane. The AS probe exhibits a fluorescence emission peak corresponding to a hexane-like environment, while that of ANS corresponds to an 80:20 mixture of alcohol and water[12]. Thus AS is located in a low dielectric region of the membrane appropriate to the methylene or terminal methyl groups of the lipid hydrocarbon, and ANS in a high dielectric region such as that of the phosphate headgroups of the hydrocarbon chains.

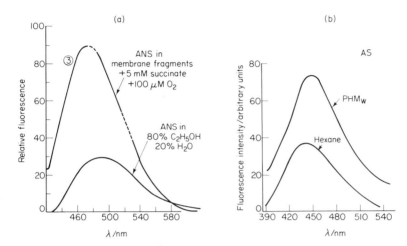

Figure 1.2 Illustrating the effect of the membrane environment upon the fluorescence emission spectra of a probe of the aqueous interface (ANS) and a probe of the hydrocarbon phase (AS). These may be compared with the emission spectra of the same probes in, respectively, an 80:20 alcohol:water mixture and in hexane. (From Chance[12], by courtesy of the USSR Academy of Sciences.)

The merocyanine probes show a similar but relatively smaller response to solvent dielectric constant; their emission spectra in the membrane is best matched by their spectra in benzene[21]. Instead of a blue shift of fluorescence emission, the merocyanines exhibit an intensification of their fluorescence. Their absorption spectrum is also sensitive to the probe environment. (the progression of spectral shifts caused by the binding of MC-V to submitochondrial particles is shown in more detail in Figure 1.7 below).

Thus the fluorescence emission wavelengths of ANS and AS, and the intensity of the fluorescence emission and the absorption wavelength of the merocyanines, may be used to provide an approximate location of these probes in the membrane. A second approach is provided by fluorescence quenching by oxidised ubiquinone.

1.5.2 Fluorescence quenching by ubiquinone

The collisional quenching of the fluorescence of ANS, AS and MC-V by the ubiquinone component affords a second and somewhat more accurate approach to the localisation of the probes in the membrane. Accepting the conclusion, based largely on solvent extraction studies, that ubiquinone occupies the hydrocarbon phase of the membrane[13], collisional quenching of a probe by ubiquinone would identify similar occupancy sites for the two molecules. The fluorescence of AS and MC-V is indeed quenched by ubiquinone but that of ANS is not; these results are consistent with the location of the ANS and AS probes as determined by x-ray and n.m.r. studies (see below), and suggest that any fluorescent probe which penetrates the membrane beyond the ester bond region is likely to be quenched by ubiquinone. On the basis of this determination, a variety of probes in which the distance between the charged group and the chromophore varies from 15 to 5 Å show progressively decreasing unbiquinone quenching as the distance between the two is decreased. This result can again be interpreted in accordance with the data on ANS and AS, namely, that ubiquinone quenching inside the ester bond region of the membrane is possible.

More recent data on the location of the probe 2-(9-anthroyl)palmitate (AP) (G. K. Radda, personal communication) indicate that it penetrates just to the beginning of the hydrocarbon layer, i.e. to the first methylene group of the hydrocarbon chain. AP is, however, effectively quenched by ubiquinone, thus identifying the range over which ubiquinone quenching can be expected to locate probes of the hydrocarbon phase as extending from the first methylene group for AP inwards 15 Å to the methylene group region in the case of AS[15].

Fluorescence quenching by resonance energy transfer, discussed above (Section 1.4.2.3) occurs over much larger distances and is more useful for determining the location of probes in the plane of the membrane rather than perpendicular to it. The more sophisticated methods outlined below are required for a more exact localisation of the probes in the membrane.

1.6 PROBE LOCALISATION BY MAGNETIC RESONANCE SPECTROSCOPY

If one is willing to accept a slower readout of probe location in the membrane and at the same time a restriction of the approach mainly to artificial lipid vesicles, then n.m.r. spectroscopy has a number of advantages over optical methods for probe localisation. Two approaches have been found to be particularly effective.

In the first case, the aromatic portions of the probe cause a shift of lipid protons in regions of closest approach of the probe molecule that is in the same direction as that caused by ferricyanide (cf. Figure 1.1); this is an upfield shift. The aromatic ring of ANS or the anthroyl group of AS are particularly effective; these are termed 'shift probes' and perturb the protons in the environment closest to the chromophore. Merocyanines have less aromaticity than do ANS and AS, but have been found to cause sufficiently measurable ring current shifts that they may ultimately provide a precise location.

Alternatively, a paramagnetic component may be used to affect the protons of the probe, i.e. the membrane component acts as a probe for the fluorescent

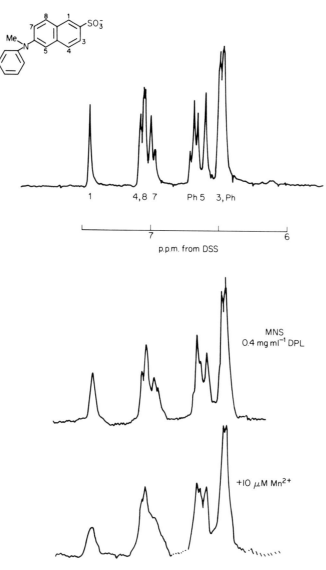

Figure 1.3 Illustrating the selective effects of membrane-bound Mn^{2+} upon the linewidths of the proton resonances (250 MHz) of MNS (5 mM) when Mn^{2+} (10 μM) is bound to the phosphate headgroup region of the dipalmitoyl-lecithin vesicles. The principal effects are upon protons 1 and 5 and the methyl protons. (From Barker et al.[52], by courtesy of Elsevier.)

molecule. This approach requires the introduction of a paramagnetic centre into the membrane, such as Mn^{2+} as used previously by the reviewer[12] and by Barker *et al.*[52], or possibly gadolinium (*cf.* Case and Leigh[50]). Several natural subdivisions of this approach occur[53]. In studies[12] with Hershberg, paramagnetic labelling of the phosphate headgroups of the membrane lipid with Mn^{2+} and determination of the broadening of the various proton resonances of ANS indicated that the proton *ortho* to the sulphonic acid group of ANS was closest to the Mn^{2+}. As shown in Figure 1.3, this method has been subsequently used by Radda and his colleagues[52] to determine the location of 2-(*N*-methylanilino)naphthalene-6-sulphonate (MNS) in lecithin vesicles. Here, the 1, 5, and methyl protons are principally affected, and these data, together with measurements of T_1, permit a placement of MNS in a membrane region similar to that obtained for ANS (*cf.* Figure 1.4). However, MNS is turned at $90°$ with respect to the orientation proposed for ANS (cf. Figure 1.3).

In natural membranes, as opposed to artificial lipid vesicles, there is not sufficient resolution of the lipid protons to identify the structural features of the lipid; even with ^{13}C n.m.r., difficulties are encountered in recognising individual species of lipids or portions of the lipid structure. The reason for this is a considerable broadening of the proton and carbon resonances in the natural membrane over those in the artificial membrane. Thus, the only technique that remains applicable to the intact membranes is based on the interaction of paramagnetic ions with the e.s.r. signal of an e.s.r. probe or of a paramagnetic membrane protein. With this approach, the localised, one-dimensional distance measurement can be relatively accurate, but the over-view of the entire structure presented by n.m.r. is missing. The following paragraphs afford examples of chemical shifts of lipid protons by two extrinsic probes and one endogenous constituent of electron transport membranes.

1.6.1   1-Anilinonaphthalene-8-sulphonate (ANS)

Fourier transform[54] or continuous-wave n.m.r. afford a measurement of the chemical shifts of the lecithin protons in single-walled vesicles owing to the introduction of ANS. These observations, in contrast to the x-ray studies reported below, are made above the transition temperature. It is customary in such experiments to obtain as a calibration point the proton resonance of the methyl group of acetate, and the chemical shifts, measured in p.p.m., are based upon this peak as reference. A positive chemical shift corresponds to a higher field resonance and the peak assignment is based upon data for lecithin in water and chloroform[55] and upon work done by Chan and his colleagues[56].

Figure 1.4 summarises the n.m.r. data on the use of ANS as a shift probe of the lipid protons and of Mn^{2+} as a perturbant of the H^{-2} and H^{-4} protons of ANS, together with data based on x-ray and optical transition moments for locating and orienting ANS between the phosphate headgroups of the lipid[12]. The upfield chemical shifts are indicated on the diagram below the particular protons of the lipid, particularly *N*-choline methyl (+0.19), methylene (+0.21) and the first methylene group of the fatty acid chain (+0.15). At the terminal methyl end of the chain there are no significant shifts, and the shifts are just detectable in the methylene groups of the fatty acid chains. The second methylene resonance from the phosphate head group has been studied by graphical analysis of the overlapping methylene resonances of the $(CH_2)_n$ band, and essentially no shift is involved. The upfield shifts are due to circulating currents in the aromatic rings of the ANS molecule 'ring currents'. The localisation of ANS is based upon the particular protons shifted since the shifts occur only at small distances between the circulating ring currents and the protons of the lipid; in fact, the shift (which

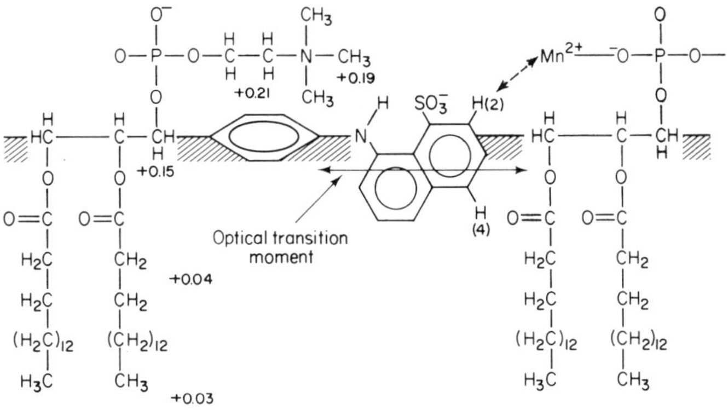

Figure 1.4 Location of ANS in DPL vesicles according to n.m.r., x-ray, transition moment and fluorescent probe data. (From Chance[12], by courtesy of the USSR Academy of Sciences.)

is upfield for these particular conditions) is localised in a conical region whose axis is perpendicular to the aromatic ring and is appreciable at distances of only a few Å. The magnitude of the effect is proportional to the magnitude of the ring currents and varies with the inverse sixth power of the distance between the aromatic ring and the lipid protons.

The figure gives a location of ANS conforming to these data. The difficult question of whether the aromatic group causing the ring-current shifts is the anilino or the naphthalene group is resolved primarily by the selective effects of Mn^{2+} bound to phosphate upon the H-2 naphthalene protons, and by optical data which locate the transition moment of the naphthalene ring parallel to the plane of the bilayer. Clearly, MNS is oriented differently than ANS (*cf.* Figure 1.3).

1.6.2 12-(9-Anthroyl)stearate (AS)

Again above the melting point, the addition of AS to dipalmitoyl-lecithin (DPL) vesicles at a ratio of 1:4 causes asymmetry in the upfield side of the methylene methyl band, as shown in Figure 1.5. Although the asymmetry is small, a graphical analysis of the area of the upfield component of the main

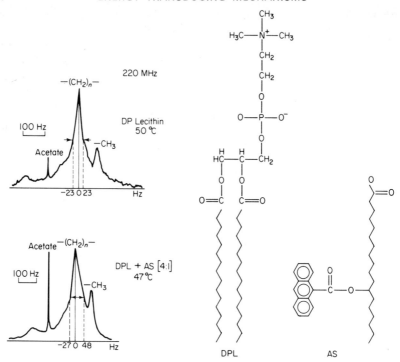

Figure 1.5 Location of AS in the membrane. (From Chance[12], by courtesy of the USSR Academy of Sciences.)

$(CH_2)_n$ band corresponds to four methyl groups shifted upfield about 0.2 p.p.m. This upfield shift, again interpreted as being due to a ring-current effect induced by the anthroyl group of the probe, locates this group opposite the methylene groups and thus about 15 Å from the edge of the bilayer.

1.6.3 Merocyanine-V

The method has not yet been applied in detail to the merocyanine dye, MC-V. The chemical shift for the merocyanine-occupied membrane is smaller with respect to the unoccupied membrane than with other probes, owing in large part to the character of the aromatic groups of the merocyanine; for example, the addition of an anthroyl group to MC-V might afford larger shifts for n.m.r. work.

1.6.4 Ubiquinone

The effect of ubiquinone upon the lipid protons has recently been studied by Podo*. The proton resonances of the methylene groups of the lipid have been identified and between these two are located the resonances of the

* Podo, F. and Blasie, J. K. (1975). Personal communication.

methylene groups of the oxidised quinone. This is resolved to some extent even in the membrane to show the terminal methyl, the methylene and the quinoid methyl protons. So far, the only interaction of oxidised ubiquinone with the lipid protons is shown as a slight broadening of the methylene group resonances of the hydrocarbon chains. Thus, ubiquinone itself is not a good shift probe and the n.m.r. studies do not describe the exact location of this component in the lipid[12].

1.6.5 Summary

These four examples differ in important ways. ANS is a sufficiently effective shift probe that it can be located unambiguously in the membrane by n.m.r. AS is less effective, but affords a small asymmetry of considerable significance in identifying its location. Merocyanine and ubiquinone are relatively ineffective and the chemical shift technique is not as useful for these probes.

Other approaches may, however, be used. In cases where the probe exchanges rapidly with the membrane, the binding of a paramagnetic ion to the phosphate headgroup of the membrane lipid affords a precise identification of the portion of the probe that approaches most closely, as noted above for ANS[12] and MNS[49]. For probes which are tightly bound to the membrane and not in rapid exchange, the chemical shift method must be used, and probes designed to give not only an environmentally responsive fluorescence signal but also a large chemical shift of the lipid protons would seem to be appropriate design goals.

1.7 PROBE LOCALISATION BY X-RAY DETERMINATIONS

In the slowest time range are the x-ray diffraction data, usually derived from oriented hydrated lecithin multilayers containing fluorescent probe molecules. There are, however, some constraints upon this method.

1.7.1 Limitations of the x-ray technique

1.7.1.1 Orientation of the membranes

The first constraint has already been mentioned: the better the degree of orientation of the membrane, the simpler the interpretation and the shorter the time for data acquisition[58]. Such orientation can be obtained in several ways. An ideal natural material is afforded by the retinal membrane protein rhodopsin, where the degree of orientation of the natural membrane is high because of the parallel structure of the outer segments and the disk membranes. Otherwise, oriented multilayers may be obtained by the Langmuir–Blodgett technique[59].

1.7.1.2 Phase of the membrane

A second fundamental problem in x-ray studies is that the membrane should be in the 'frozen' rather than the 'melted' state[15, 60]. Although this is not essential, the interpretation of the structure of the fatty acid chains will be diminished by blurring of the diffraction data owing to thermal disordering of the terminal methyl groups of the hydrocarbon chains of the lipid. This is illustrated in Figure 1.6a, which shows the electron density profiles for DPL

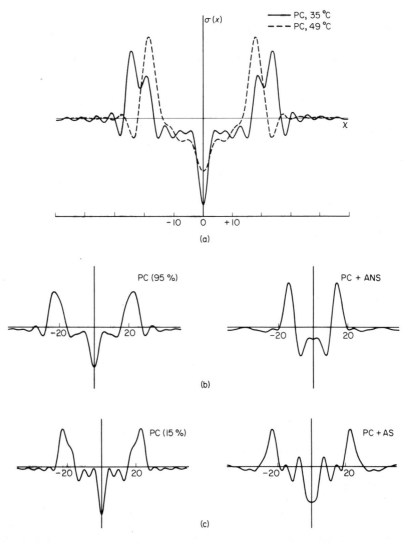

Figure 1.6 (a) Electron density profiles for dipalmitoyl-lecithin bilayers above and below the phase transition temperature; (b) electron density profiles for DPL and DPL:ANS = 3:2 bilayer; (c) electron density profiles for DPL and DPL:AS = 4:1 bilayers. (From Chance[12], by courtesy of the USSR Academy of Sciences.)

multilayers above 49 °C (dashed curve) and below 35 °C (solid curve) the melting point of the lipid. An interpretation of the electron density profile based upon the chemical structure of the components of the bilayer, as well as on the x-ray diffraction data arising from the packing of the chain, indicates that the outermost peaks correspond to the phosphate and trimethyl-ammonium groups which lie in the same plane. The secondary, inner shoulder corresponds to the ester bonds of the lipid, which also lie in the same plane. The fatty acid chains are then extended chiefly in an all-*trans* configuration normal to the plane of the bilayer, and are packed in a hexagonal lattice as well, except for the last few carbon atoms which show thermal disorder and thus give the central trough.

Melting of the bilayer causes a highly significant thinning (the distance between the peaks is about 40 Å) and the ester bond region is not resolved, presumably owing to thermal motion of the components of this portion of the membrane. The lipid chains are now no longer confined to the hexagonal lattice and appear to have less of the all-*trans* configuration[61] and more of the *gauche* $\pm$ configuration about single bonds in the hydrocarbon region. Nevertheless, the terminal methyl groups remain localised predominantly in the centre of the bilayer, and thermal disorder continues to give the central trough of the profile as before. For these reasons, the x-ray data are usually taken below the melting point in order better to resolve the structural changes induced in the lipid bilayer by the presence of the probe molecule.

1.7.2   1-Anilinonaphthalene-8-sulphonate (ANS)

An example of conditions appropriate for x-ray studies is afforded by the electron density diagram of Figure 1.6b, which shows ANS-occupied DPL bilayers at a ratio of DPL:ANS = 3:2. The diffraction pattern was taken below T_M, but nevertheless has some features similar to the diagram taken above T_M; the bilayer is thinner and the characteristic shoulder of the ester bond region has disappeared. However, the central trough is much larger in the occupied than in the unoccupied membrane, suggesting an even greater disordering of the hexagonal lattice in the plane of the bilayer. In fact, interdigitation of the fatty acid chains from each of the two opposed monolayers is proposed as an explanation of the flattening of the trough in the electron density profile for the centre of the bilayer[62].

1.7.3   12-(9-Anthroyl)stearate (AS)

If, instead of ANS, AS occupies the lipid bilayer in a ratio of DPL:AS = 4:1, the thinning of the bilayer is scarcely noticeable, as seen in Figure 1.6c[12]. However, the ester bond region is no longer resolvable, as it was in the case of ANS. The main feature is a large increase in the width of the central trough of the profile, suggesting again a statistically disordered state of the lipid chains in the plane of the bilayer. Thus, AS appears to be in a relatively extended form, with its carbonyl group near the phosphate head-groups and its anthroyl moiety within the hydrocarbon core, thereby disturbing the chain packing[15]. It is of interest to note that this perturbation does not occur with the incorporation of stearic acid alone in the same ratio to lipid.

1.7.4   Ubiquinone

Cain et al.[62] locate oxidised ubiquinone in DPL bilayers below the transition temperature with the quinone moiety in the ester bond region. The branched unsaturated hydrocarbon chain of the quinone penetrates in an extended form to the hydrocarbon layer of the bilayer. It is possible that hydrogen bonding between the reduced ubiquinone protons and the ester-bond region of the lipid is the cause of more regular chain packing, presumably causing the difference in the collisional effectiveness of the oxidised and reduced forms of ubiquinone in quenching the AS fluorescence. However, the oxidised and reduced forms show different quenching of AS fluorescence in solution, and for this reason the differences in the lipid may not be attributable strictly to structural phenomena[12].

1.7.5   Merocyanines

At present, x-ray diffraction data for the lipid bilayer occupied by the merocyanine probes are not available.

1.7.6   Summary

The effects of all these perturbations of the basic x-ray diffraction pattern of 'frozen' dipalmitoyl-lecithin oriented multilayers are different. The main feature common to all three perturbations is the loss of detail in the region of the phosphate head-groups; the ester-bond region can no longer be determined in any of the three states. Thinning of the bilayer occurs in the presence of ANS. Deepening of the central trough is most pronounced with AS, although a shallowing of this trough is observed with ANS; the latter effect, however, is a consequence of interdigitation of the lipid chains from opposite sides of the bilayer.

1.8   ELECTROCHROMIC RESPONSES OF PROBES

Various cyanine dyes, such as the carbocyanines employed by Bücher et al.[26], or the merocyanines used by Cohen et al.[29,63-65] in the squid axon and by the reviewer in chromatophores[21] and submitochondrial particles[22], and described in detail by Sims et al.[66], are members of a class of dyes capable of reporting changes of membrane charge by both fluorescence and absorption changes.

1.8.1   Theoretical considerations

An electrochromic effect, as defined by Platt[23,24] and interpreted here (cf. also the appendix to Ref. 21), refers to a spectral shift of the absorption band or of the fluorescence emission band of a given molecule owing to the presence of an electric field, and is essentially a 'Stark effect'. The two principal terms of the equation below vary with the field, one linearly and the other quadratically:

$$h\Delta v = -\Delta\mu\cdot F - \tfrac{1}{2}\Delta aF^2 \qquad (1.1)$$

where $h\Delta v$, of course, is the wavelength shift, μ is the permanent dipole moment, a is the induced dipole moment and F is the magnitude of the field.

In the first term, $-\Delta\mu\cdot F$, the sign of the effect will vary with the sign of the field; in the second term, $\tfrac{1}{2}\Delta aF^2$, it will not. The first and second terms may be isolated from each other in two ways. In experiments by Bücher and his colleagues[26], and by Schmidt and Reich[27], the field, F, is varied sinusoidally and the first and second harmonics of $h\Delta v$ are observed. These amplitudes correspond to the first and second terms of the equation, and therefore to the appropriate coefficients. Alternatively, as recently pointed out by Cheng (personal communication), the first derivative of the equation with respect to F gives a straight line with the coefficient of the first term as its intercept and the coefficient of the second term as its slope.

The coefficients of the equation are of interest in themselves. The relationship between the orientation of $\Delta\mu$, the permanent dipole moment, and F, as well as the change of permanent dipole moment. $\Delta\mu$, from the excited to the ground state ($\Delta\mu_e \rightarrow \Delta\mu_g$) are of fundamental importance. Obviously, if $\Delta\mu$ is zero due either to a lack of change in μ between the excited and ground states, or to the fact that the changes in the ground and excited states are of the same magnitude but opposite sign, the first term would be nullified.

Orientation effects, for example μ being perpendicular to F, or orientations of the excited and ground state values of μ such that $\Delta\mu = 0$, could also nullify the first term. Indeed, these may be just the phenomena which caused Schmidt and Reich[27] to fail to obtain a linear term in the electrochromism of the carotenoid lutein. In other cases where $\Delta\mu$ is finite and oriented appropriately to F, one will have a positive (blue) or negative (red) shift, depending upon the orientation of $\Delta\mu$ and F, and upon the sense of F.

Similar limitations apply to the quadratic shift, namely that Δa be finite. However, since F is in a squared term, the orientation problems are much less severe for quadratic than for linear effects. The sign of the frequency shift depends upon the sign of Δa. One orientation restriction remains, however; the probe molecule must be orientated appropriately so that the exciting photon is absorbed.

The importance of analysing the quadratic response cannot be underestimated, since Schmidt et al.[67] found only a quadratic electrochromism in the carotenoid lutein, which can be assumed to be a representative example of carotenoids in chloroplasts and chromatophores. However, in a particular case of an Rps. spheroides mutant containing primarily neurosporene as carotenoid, Crofts[68] found red shifts induced by diffusion potentials that he considers to be linear effects; however, they may well have been quadratic effects (see below).

From these studies and from the study of the cyanine dye by Bücher[26], it appears that quadratic electrochromism is more likely and that linear electrochromism may not be observed experimentally under the conditions listed above. Which one of these conditions leads to the lack of linear electrochromism in lutein is not clear, but it is probable that orientation factors are not involved, and either the first or second condition applies, namely, that $\mu_e = \mu_g = 0$, or $\mu_e = \mu_g$, rendering $\Delta\mu = 0$. The experimental observation of a linear relation between the diffusion potential and the red shift of the carotenoid is difficult to explain by the suggestion of Witt that a large local field exists which shifts the carotenoid on to the linear portion of the quadratic characteristic of red shift vs. potential, since no large electrochromic shift has been reported upon extraction or reincorporation of carotenoid[67].

The systematic study of the effect of localised and delocalised potentials upon the probe responses has yet to be carried out over a range sufficient to identify clearly the coefficients of the linear and quadratic terms. Attempts to establish a diffusion potential are often frustrated by the experimental difficulties of obtaining ion-free vesicles on the low concentration side, or, on the high concentration side, sufficient salt concentrations without optical artifacts. The possibility of calibrating these probes by establishing potentials across a lipid bilayer or large single-walled vesicle is attractive, but has yet to be explored in detail. The information ultimately to be gained from such studies is of fundamental importance; the identification of the two coefficients of the general equation for the frequency shift leads to the quantitation of the probe parameters in terms of the permanent and induced dipole moments. A suitable approach would appear to be a direct one, in which the potential is applied directly to a multilayer, as has been done by Bücher et al.[26] and by Schmidt and Reich[27]. The interpretability of the data will be much greater if such results can be obtained for the bilayer where the portion of the total potential applied to the probe is much better defined.

The experimental approaches to the calibration of voltage-sensitive probes will not be further discussed here, since much still remains to be done to extend the work already in hand and to improve the technology of the calibration procedure. Instead, this review will be concluded with some examples of energy-dependent responses of merocyanine probes in natural systems that are consistent with the equation above for electrochromic responses. Two examples are relevant: that afforded by chromatophores where energisation is light-induced, and that observed in submitochondrial particles where the energisation is by electron flow to oxygen with oxygen reduction. Other examples where the merocyanine probes are employed as indicators of membrane potential will be only briefly indicated here, based on results kindly communicated by Drs. L. B. Cohen, B. M. Salzberg, H. V. Davila and W. N. Ross.

1.8.2 Electrochromic responses in natural systems

1.8.2.1 Red shifts of merocyanine-V absorption

(a) *Submitochondrial particles*—Taking first membrane energisation by electron transfer, Figure 1.7 shows the absorption spectrum of the membrane-bound probe as trace 1, which does not contain the absorption of the chromophores of the vesicles since the computer attached to the spectrophotometer

memorised these absorption bands and they were subsequently read out as a
correction to the measuring wavelength for these specific absorbancies. The
absorption of the probe was not, however, compensated and therefore
appears as a spectrum peaking at 628 nm (see trace 1). The submitochondrial
particles are supplemented with oligomycin and are in a Tris-sulphate buffer
of low ionic strength, under which conditions other ions do not interfere with
the probe response. In trace 2 at 77 s per scan, electron transport initiated
by the addition of 5 mM succinate causes a rapid red shift of 4 nm, the
new peak being at 632 nm. Subsequent scans do not enhance the red shift
but show a small intensification of the peak at 632 nm and a loss of
absorption in the region of 600 nm. This biphasic response of the merocyanine probes in submitochondrial particles is observed to a much greater
extent with MC-I and MC-II[22]. The response is reversible; an addition of
valinomycin and K^+ or of uncoupling agents will reverse the effect, rapidly
for the red shift and more slowly for the effect at 600 nm.

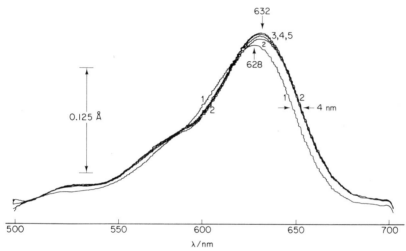

Figure 1.7 Red shifts of merocyanine-V absorption caused by energisation in biological membranes: by succinate-induced electron transport in submitochondrial
particles

(b) *Chromatophores*—A similar result is obtained with probe-supplemented
chromatophores, where activation of electron transport is caused by infrared
illumination. Thus the control traces 1 and 2 of Figure 1.8 are included to
ensure that no spectroscopic changes occur in the region of probe absorption
before the probe is added. Trace 1 represents the baseline derived from
memorisation of the absorption of the pigments of the chromatophores;
deletion of the absorption bands leaves a straight baseline.

Illumination of the chromatophores at 860 nm causes shifts in the region
of the carotenoid bands (480–530 nm) and small changes in the region of
bacterial chlorophyll (580–630 nm). Addition of the probe in trace 3 gives a
peak at approximately 618 nm, similar to that obtained in the submitochondrial particles of Figure 1.7. In trace 4, illumination of the chromatophores with 860 nm light of the same intensity as that used in trace 2 now

causes, in addition to similar changes in the carotenoid and chlorophyll regions, a large shift of the probe spectrum towards the red to the extent of 5.5 nm, bringing the peak to 624 nm. As in the case of the submitochondrial particles, the red shift is readily reversible by cessation of illumination and by uncoupling agents in the illuminated chromatophores.

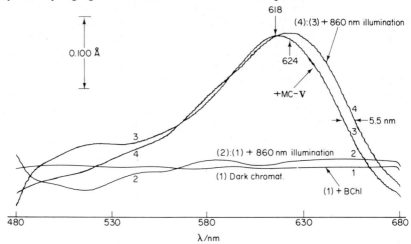

Figure 1.8 Red shifts of merocyanine-V absorption caused by energisation in biological membranes: by light-induced electron transport in chromatophores of *R. rubrum*

(c) *Interpretation*—These two examples indicate that a red shift of the absorption of the merocyanine probe is readily obtained upon membrane energisation*. The interpretation of the red shift as linear or quadratic, and the values of $\Delta\mu$ and Δa have not yet been determined simply because of the difficulty of identifying *in situ* the membrane transition that affords a continuous linear variation of localised or delocalised charge. Furthermore, experiments with model membrane systems, especially lipid bilayers in which a transmembrane field can be established, have yet to be carried out with this type of probe. Thus, the main interpretation of the response of the probe absorption is based on the spectra afforded by Bücher *et al.*[26] in their study of the carbocyanine probe in which the red shift was identified as a quadratic response for this dye, whose structure is closely related to that of the merocyanines.

1.8.2.2 *Changes of merocyanine-V fluorescence*

Figure 1.9 bridges the gap between the red shift of the previous figures and the fluorescence change. R. rubrum chromatophores—capable of energy *coupling*[69]—are supplemented with MC-V and illuminated at 860 nm. The abrupt downward deflection of the trace indicates the time course of the red shift, as indicated by Figure 1.8. After 50 s of illumination the light is turned off and the red shift decays with a half-time of about 10 s.

* Under our conditions no other significant absorption changes are observed, such as those reported for merocyanine I in the squid axon[64].

At the same time as the fluorescence change is recorded an abrupt decrease and a slow increase occurs, apparently on the same timescale as the red shift. Thus, membrane energisation and the associated hydrogen ion binding to the membrane[70] result in a fluorescence decrease. Similar results are obtained upon energisation of chromatophores either by adding succinate to the substrate-free particles or oxygen to the anaerobic substrate supplemented system.

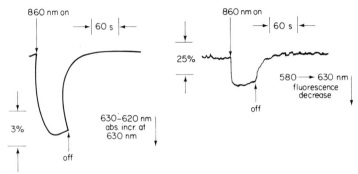

Figure 1.9   Absorbance and fluorescence changes caused by energisation of *R. rubrum* chromatophores upon i.r. illumination. 16 μM bacterial chlorophyll *R. rubrum* chromatophores, 4.1 μM MC-V, 5 mM tris-sulphate, pH 7.2

Thus changes of fluorescence in the membrane energisation are observed in submitochondrial particles[21], chromatophores[22], and in ionophore-induced K^+ influx into these membranes as well as into artificial lipid vesicles (unpublished work with J. Smith, G. Cheng, and P. Mueller). These results are summarised in Table 1.1. On the bottom lines of the table are indicated the red shifts (upward arrows) and the fluorescence decreases (downward arrow) observed in energisation of submitochondrial particles and chromatophores (see Figure 1.9). Calibrations of the sign of the fluorescence change in these two systems, and in lecithin vesicles as well, are included in the first two lines, and the consistent result for the three systems is a fluorescence decrease due to K^+ efflux. Such changes can be observed in the neurons and glia of the rat brain cortex when massive K^+ efflux occurs during the course of spreading depression[71].

A correlation of these changes with changes of membrane potential is afforded by the illustration in Figure 1.10[72]. Here the decrease of potential is caused by an internal electrode driven by a voltage 'clamp'. When the electrode is made positive, the membrane potential is decreased and depolarisation occurs with the corresponding increase of fluorescence; the opposite occurs when the electrode potential is decreased, i.e. made more negative, and hyperpolarisation occurs leading to decreased fluorescence. The fact that the amplitude of the response for both depolarisation and for hyperpolarisation are approximately equal suggests that the fluorescence changes have a linear dependence upon the potential. The magnitude of the fluorescence change obtained with 600 nm excitation and the emission measurement of about 630 nm amounts to 2.5×10^{-5} of the initial signal, less than that obtained with the M-I probe, in the squid axon[73].

Table 1.1 Sign of merocyanine V (MC-V) changes in various membrane systems*

| | Lecithin vesicles | | Submitochondrial particles | | Chromatophores (R.r.) | | Rat brain cortex | Squid axon |
	A	F	A	F	A	F	F	F
K⁺ influx	←	←	←	←	←	←--	←	
K⁺ efflux		→		→	→	→	→	
Depolarisation								←
Hyperpolarisation								→
Energisation			←	→	←	→		

* ↑ denotes a real shift or a fluorescence increase.

The first important observation is that the sign and magnitude of the fluorescence change follow the sign and magnitude of the applied potential. Thus, linear electrochromism involving the term $\Delta\mu \cdot F$ is observed. Since the magnitudes are identical to within the experimental error, there is no evidence for an additional quadratic electrochromism in the fluorescence change. The response time in the two traces is less than 1 ms but differs slightly for the two polarities, owing to an incomplete compensation for the series resistance (*cf.* figure 8 of Ref. 70).

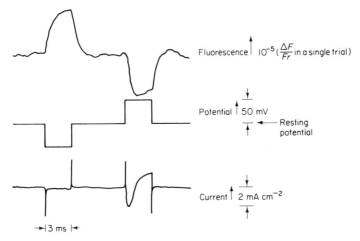

Fluorescence 10^{-5} ($\frac{\Delta F}{Fr}$ in a single trial)

Potential 50 mV
Resting potential

Current 2 mA cm^{-2}

3 ms

Figure 1.10 Response of MC-V to depolarisation and hyperpolarisation of a squid axon. (Reproduced by courtesy of Dr. L. B. Cohen *et al.*)

If one then takes these data on fluorescence change, where only a linear electrochromism is obtained, and applies Bücher's data[26] to the absorbancy changes in submitochondrial particles and chromatophores which are interpreted as a red shift [equation (1.1)], it may be concluded that in the fluorescence change the first term of equation (1.1), $\Delta\mu \cdot F$, is governing. If this concept is confirmed in further tests on model membranes, one may have a situation in which linear and quadratic electrochromism are separately measured by fluorescence and absorption measurements of the same probe molecule, leading to a direct determination of the two parameters, $\Delta\mu$ and Δa, from the linear first derivative plots of equation (1.1).

Finally, the increments of membrane potential obtained by the two methods must eventually be calibrated for the linear or quadratic components of the fluorescence and absorption changes.

Acknowledgement

The preparation of this review was supported by grants USPHS GM-12202 and NINDS 10939.

References

1. Podo, F. and Blasie, J. K. (1975). In *MTP International Review of Science, Biochemistry Series One*, Vol. 2, *Biochemistry of Cell Walls and Membranes* (G. F. Fox, editor) (London: Butterworths)
2. Radda, G. K. and Vanderkooi, J. (1972). *Biochim. Biophys. Acta*, **265**, 509
3. Jost, P., Waggoner, A. S. and Griffith, O. H. (1971). *Structure and Function of Biological Membranes*, 83 (L. S. Rothfield, editor) (New York: Academic Press)
4. Steim, J. M., Tourtelotte, M. E., Reinert, J. C., McElhaney, R. N. and Rader, R. L. (1969). *Proc. Nat. Acad. Sci. USA*, **63**, 104
5. Kirk, R. G. and Tosteson, D. C. (1973). *J. Membrane Biol.*, **12**, 273
6. Packer, L., Donovan, M. P. and Wrigglesworth, J. M. (1971). *Probes of Structure and Function of Macromolecules and Membranes. I. Probes and Membrane Function*, 219 (B. Chance, C. P. Lee and J. K. Blasie, editors) (New York: Academic Press)
7. Hackenbrock, C. R. (1966). *J. Cell Biol.*, **30**, 269
8. Hackenbrock, C. R. (1968). *J. Cell Biol.*, **37**, 345
9. Green, D. E., Asai, J., Harris, R. A. and Penniston, J. T. (1968). *Arch. Biochem. Biophys*, **125**, 684
10. Chance, B., Azzi, A., Lee, I. Y., Lee, C. P. and Mela, L. (1969). *Mitochondria: Structure and Function, FEBS Symp.*, Vol. 17, 233 (L. Ernster and Z. Drahota, editors) (London: Academic Press)
11. Waggoner, A. S. and Stryer, L. (1970). *Proc. Nat. Acad. Sci. USA*, **67**, 579
12. Chance, B. (1972). *Proc. Fourth Int. Congr. Biophys.*, 911 (L. Kayushin, editor) (Moscow: USSR Academy of Sciences)
13. Chance, B. (1972). *Biochemistry and Biophysics of Mitochondrial Membranes*, 85 (G. F. Azzone, E. Carafoli, A. L. Lehninger, E. Quagliariello and N. Siliprandi, editors) (New York: Academic Press)
14. Weber, G. and Laurence, D. J. R. (1954). *Biochem. J.*, **56**, 31P
15. Lesslauer, W., Cain, J. E. and Blasie, J. K. (1972). *Proc. Nat. Acad. Sci. USA*, **67**, 499
16. Hoffman, J. F. and Laris, P. C. (1974). *J. Physiol. (London)*, in the press
17. Azzi, A., Chance, B., Radda, G. K. and Lee, C. P. (1969). *Proc. Nat. Acad. Sci. USA*, **62**, 612
18. Azzi, A. (1969). *Biochem. Biophys. Res. Commun.*, **37**, 254
19. Brocklehurst, J. R., Freedman, R. B., Hancock, D. J. and Radda, G. K. (1970). *Biochem. J.*, **116**, 721
20. Azzi, A. and Santato, M. (1971). *Biochem. Biophys. Res. Commun.*, **44**, 211
21. Chance, B. and Baltscheffsky, M. (1975). *Biomembranes*, 33 (L. Manson, editor) (New York: Plenum.)
22. Chance, B. and Baltscheffsky, M. (1974). *Perspectives in Membrane Biology*, 329 (S. Estrade and C. Gitler, editors) (New York: Academic Press)
23. Platt, J. R. (1956). *J. Chem. Phys.*, **25**, 80
24. Platt, J. R. (1961). *J. Chem. Phys.*, **34**, 862
25. Brooker, L. G. S., Cragi, A. C., Heseltine, D. W., Jenkins, P. W. and Lincoln, L. L. (1965). *J. Amer. Chem. Soc.*, **87**, 2443
26. Bücher, H., Weigand, J., Snavely, B. B., Beck, K. H. and Kuhn, H. (1969). *Chem. Phys. Lett.*, **3**, 508
27. Schmidt, S. and Reich, R. (1972). *Ber. Bunsenges. Phys. Chem.*, **76**, 589, 599, 1202
28. Netzel, T. L., Rentzipes, P. M. and Leigh, J. S. (1973). *Science*, **182**, 238
29. Cohen, L. B., Salzberg, B. M., Davila, H. V., Ross, W. N., Landowne, D., Waggoner, A. S. and Wang, C. H. (1974). *J. Memb. Biol.*, in the press
30. Vanderkooi, J. M. and Callis, J. (1974). *Biochemistry*, in the press
31. Hubbell, W. L. and McConnell, H. M. (1971). *J. Amer. Chem. Soc.*, **93**, 314
32. Thomas, D. D. and McConnell, H. M. (1974). *Chem. Phys. Lett.*, **25**, 470
33. Magde, D., Elson, E. L. and Webb, W. W. (1974). *Biopolymers*, **13**, 29
34. Cone, R. A. (1972). *Nature New Biol.*, **236**, 39
35. Junge, W. and Eckhof, A. (1973). *FEBS Lett.*, **36**, 207
36. Junge, W. (1972). *FEBS Lett.*, **25**, 109
37. Chance, B. and Erecinska, M. (1971). *Arch. Biochem. Biophys.*, **143**, 675

38. Hoenig, H. E., Wang, R. H., Rossman, G. R. and Mercereau, J. E. (1972). *Proc. Applied Superconductivity Conference*, 570 (New York: IEEE)
39. Brill, A. S. (1956). *Development of a Fast and Sensitive Magnetic Susceptometer for the Study of Rapid Biochemical Reactions* (Ph.D. Thesis, University of Pennsylvania)
40. Leigh, J. B., Holmes, K. C. and Rosenbaum, G. (1973). *Research Applications of Synchrotron Radiation* (R. E. Watson and M. Perlman, editors) (Upton: Brookhaven National Laboratory)
41. Chanbre, M. and Cavaggioni, A. (1973). *Biophys. J.*, **13**, 235A
42. Schwartz, S., Cain, J., Dratz, E. and Blasie, J. K. (1974). *Fed. Proc.*, **33**, 1575
43. Dupont, G., Gabriel, A., Chabre, M., Gulik-Krywicki, T. and Scheckter, E. (1972). *Nature*, **238**, 331
44. Weber, G. (1953). *Adv. Protein Chem.*, **8**, 416
45. Leigh, J. S. (1971). *Structural Measurements by Magnetic Resonance* (Ph.D. Thesis, University of Pennsylvania)
46. Michaelson, D. M., Horwitz, A. F. and Klein, M. P. (1973). *Biochemistry*, **12**, 2637
47. Bystrov, V. F., Dubrovina, N. I., Barsukov, L. I., and Bergelson, L. D. (1971). *Chem. Phys. Lipids*, **6**, 343
48. Slichter, C. P. (1963). *Principles of Magnetic Resonance* (New York: Harper and Row)
49. Case, G. L. (1975). *Biochim. Biophys. Acta*, **375**, 69
50. Case, G. and Leigh, J. S. (1974). *Proc. 11th Rare Earth Conf.*, 706 (H. A. Eick and J. M. Hafchke, editors) (Oakridge, Tenn.: U.S. Atomic Energy Commission)
51. Yguerabide, J. and Stryer, L. (1971). *Proc. Nat. Acad. Sci. USA*, **68**, 1217
52. Barker, R. W., Barrett-Bee, K., Berden, J. A., McCall, C. E. and Radda, G. K. (1974). *Dynamics of Energy-Transducing Membranes*, 321 (L. Ernster, R. W. Estabrook and E. C. Slater, editors) (Amsterdam: Elsevier)
53. Andrews, S. B., Faller, J. W., Gilliam, J. M. and Barrmett, R. J. (1973). *Proc. Nat. Acad. Sci. USA*, **70**, 1814
54. Farrar, T. C. and Becker, E. D. (1971). *Pulse and Fourier Transform Nuclear Magnetic Resonance Spectroscopy* (New York: Academic Press)
55. Finer, E. G., Flook, A. G. and Hauser, H. (1972). *Biochim. Biophys. Acta*, **260**, 49
56. Chan, S. I., Sheetz, M. P., Seiter, C. H. A., Feigenson, G. W., Hsu, M., Lau, A. and Yau, A. (1973). *Ann. N.Y. Acad. Sci.*, **222**, 499
57. Jackson, L. M. and Sternell, S. (1969). *Applications of Nuclear Magnetic Resonance Spectroscopy in Organic Chemistry* (Oxford: Pergamon Press)
58. Lesslauer, W. and Blasie, J. K. (1971). *Biophys. J.*, **12**, 175
59. Blodgett, K. B. (1935). *J. Amer. Chem. Soc.*, **57**, 1007
60. Lesslauer, W., Cain, J. and Blasie, J. K. (1971). *Biochim. Biophys. Acta*, **241**, 547
61. Volkenstein, M. V. (1963). *Configurational Statistics of Polymeric Chains* (*High Polymers, XVII*), 94 (New York: Interscience)
62. Cain, J., Santillan, G. and Blasie, J. K. (1972). *Membrane Research* (C. F. Fox, editor) (New York: Academic Press)
63. Cohen, L. B. (1973). *Physiol. Rev.*, **53**, 373
64. Ross, W. N., Salzberg, B. M., Cohen, L. B., and Davila, H. V. (1974). *Biophys. J.*, **14**, 983
65. Davila, H. V., Salzberg, B. M , Cohen, L. B. and Waggoner, A. S. (1973). *Nature New Biol.*, **241**, 159
66. Sims, P. J., Waggoner, A. S., Wang, C.-H. and Hoffman, J. F. (1974). *Biochemistry*, **13**, 3315
67. Schmidt, S., Reich, R. and Witt, H. J. (1972). *Proc. 2nd Int. Congr. Photosynthesis Research*, 1087 (G. Forti, M. Avron and A. Melandri, editors) (The Hague: N. W. Junk)
68. Crofts, A. R. (1974). *Perspectives in Membrane Biology*, 373 (S. Estrade and C. Gitler, editors) (New York: Academic Press)
69. Baltscheffsky, M. (1967). *Nature*, **216**, 241
70. Chance, B., Crofts, A. R., Nishimura, M. and Price, B. (1970). *Eur. J. Biochem.*, **13**, 364
71. Mayevsky, A., Zeuthen, T. and Chance, B. (1974). *Brain Res.*, **76**, 347
72. Cohen, L. B., Salzberg, B. M., Davila, H. V. and Ross, W. N. (1974). Personal communication
73. Davila, H. V., Cohen, L. B., Salzberg, B. M. and Shrivastav, B. B. (1974). *J. Membrane Biol.*, **15**, 29

2
Energy Coupling in Biological Membranes: Current State and Perspectives

V. P. SKULACHEV
Moscow State University

Abbreviations

$\Delta\mu_H$	electrochemical transmembrane potential of hydrogen ions
$\Delta\Psi$	electric transmembrane potential
CCCP	trichlorocarbonyl cyanide phenylhydrazone
DCCD	dicyclohexylcarbodi-imide
FCCP	*p*-trifluoromethoxycarbonyl cyanide phenylhydrazone
PCB$^-$	phenyldicarbaundecaborane anion
PMS	phenazine methosulphate
TPA$^+$	tetraphenylarsonium cation
TPP$^+$	tetraphenylphosphonium cation

2.1 INTRODUCTION

The history of the past seven years of oxidative phosphorylation is actually a story of competition of two alternative concepts: the old one, defined as the chemical scheme, and the new one, introduced by Mitchell[1] as the chemiosmotic hypothesis. For a long time any attempts to discriminate between these concepts failed. It is only quite recently that decisive progress in rationalising the principle of energy coupling in oxidative phosphorylation was made. The success was associated primarily with the development of methods for resolving and reconstituting the oxidative phosphorylation system and determining the formation of an electric potential by membranous enzyme complexes. In this review we would like to summarise the most important experimental findings along the above-mentioned lines and to outline the perspectives of further investigation of energy-transducing systems in biomembranes with special attention to the chemiosmotic hypothesis.

2.2 VERIFICATION OF THE CHEMIOSMOTIC HYPOTHESIS

2.2.1 The chemiosmotic principle of energy coupling

The main principle of the chemiosmotic scheme of energy coupling is that oxidation and phosphorylation are coupled via the electrochemical transmembrane potential of hydrogen ions ($\Delta\mu_H$):

$$\text{oxidation} \rightarrow \Delta\mu_H \rightarrow \text{ATP} \qquad (2.1)$$

It is this principle that keeps the chemiosmotic concept clearly apart from any other hypothesis of oxidative phosphorylation. According to equation (2.1) there are two separate enzymic mechanisms dealing with $\Delta\mu_H$ in coupling membranes: (i) a $\Delta\mu_H$-generating redox chain and (ii) $\Delta\mu_H$-utilising ATP-synthetase. Assuming the $\Delta\mu_H \rightarrow$ ATP energy transduction to be reversible, one may modify equation (2.1) as follows:

$$\text{oxidation} \rightarrow \Delta\mu_H \leftarrow \text{ATP} \qquad (2.2)$$

In other words, $\Delta\mu_H$ can be generated independently by the redox chain and by the ATPase system. To solve the energy coupling problem it is important to answer the question whether or not the above two pathways to $\Delta\mu_H$ exist in the same coupling membrane.

2.2.2 Respiratory chain and ATPase as two mechanisms of membrane potential generation in mitochondria

2.2.2.1 Methods

If there are two types of membrane potential generators (oxidative and dephosphorylative) in mitochondrial membrane, an attempt may be made to separate these two enzymic systems and to reconstitute membraneous vesicles containing oxidative or, alternatively, dephosphorylative enzymes competent in membrane potential formation. Such an attempt became possible as two methods have been developed, one for reconstituting liposomes with mitochondrial enzyme proteins incorporated into the phospholipid membrane and the other for measuring the membrane potential in vesicles even as small as liposomes.

The first method was introduced by Racker and associates[2-7]. Its principle consists of the following. Membraneous proteins isolated from mitochondria (e.g. cytochrome oxidase or hydrophobic proteins of oligomycin-sensitive ATPase) are mixed with solutions of phospholipids in cholate. The mixture is then dialysed to remove cholate. During dialysis, small closed vesicles with a $\sim$70 Å-thick membrane are formed. A chemical analysis showed that the vesicles consist of phospholipids and proteins. Hence the reconstituted particles can be defined as proteoliposomes.

To detect the membrane potential in proteoliposomes, the penetrating ion method was primarily used. This method is based on measurements of membrane potential-induced changes in penetrating ion concentrations outside particles. For example, generation of an electric potential difference across the particle membrane with *plus* inside can be revealed by uptake of an *anionic* penetrant. If there is *minus* inside, *cationic* penetrants should be taken up. To minimise the interference of any effects connected with ion-specific enzymic transport systems, a set of unnatural ions of foreign iono-phores may be used. An important requirement of the electrophoretic penetrant responses is the absolute dependence of the ion movement direction upon the orientation of the electric potential-generating process in the membrane.

Mitchell and Moyle[8] were the first to measure membrane potential in mitochondria using K^+ ions and valinomycin as a foreign ionophore. The synthetic penetrating ions as a probe for membrane potential were intro-duced and applied for different experimental systems by Liberman, the author of this review and associates[9-15]. Using this probe, we revealed mem-brane potential formation in energised submitochondrial[9, 10], subchloroplast[14] and subbacterial[12, 14] membraneous particles. In all cases an artificial phos-pholipid membrane was used as a penetrant-sensitive electrode[16].

The adequacy of the above method was confirmed by other membrane potential probes applied in this[14, 18] and other[17, 19-22] laboratories. One of the results obtained with *Rhodospirillum rubrum* chromatophores[18] is shown in Figure 2.1, which demonstrates the response of a penetrating anion, phenyl-dicarbaundecaborane (PCB⁻), electrochromic spectral shift of carotenoids and bacteriochlorophyll, fluorescence changes of atebrin responding to a ΔpH, and of anilinonaphthalene sulphonate (ANS⁻) responding to both $\Delta\Psi$ and ΔpH. It is seen that energisation of chromatophores by adding a small amount of inorganic pyrophosphate induces characteristic changes in five parameters measured. In a few minutes, reversal of all responses takes

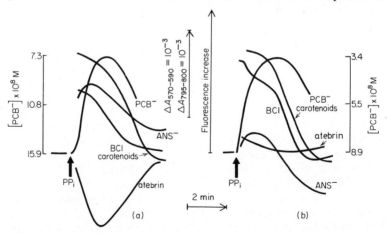

Figure 2.1 Comparison of different $\Delta\Psi$ and ΔpH probes in the *Rhodospirillum rubrum* chromatophores. Incubation mixture: 0.25 M sucrose, 0.05 M Tris-HCl (pH = 7.7), 5 mM $MgCl_2$, 0.02 M KCl, chromatophores (4×10^{-6} M chloro-phyll), concentration of ANS⁻ was 1×10^{-4} M, that of atebrin 2.5×10^{-6} M. In exp. B the mixture was supplemented with nigericin (1.5 μg ml⁻¹). Addition: 2.5×10^{-5} M sodium pyrophosphate. (From Barsky *et al.*[18], by courtesy of Elsevier.)

place owing to exhaustion of pyrophosphate (Figure 2.1a). In the presence of nigercinin, which replaces a ΔpH by ΔpK, the response of atebrin disappears, that of ANS^- decreases whereas those of PCB^- and of both electrochromic probes remain unaffected (Figure 2.1b).

Recently, a direct method for measuring electric potential difference generation in proteoliposomes was elaborated[43b]. It was found that proteoliposomes can be integrated into the planar phospholipid membrane if negative surface charges of the phospholipids are neutralised by Ca^{2+} ions. In this case, $\Delta\Psi$ generation by proteoliposome enzyme systems was shown to result in the $\Delta\Psi$ formation across planar membrane. The latter effect was measured by Ag|AgCl electrodes connected with a sensitive voltmeter. The data obtained by this method were in agreement with those of the penetrating ion experiments (see below).

2.2.2.2   Reconstitution of cytochrome oxidase generator of membrane potential

In Racker's laboratory and in our laboratory, the membrane potential generator responsible for $\Delta\Psi$ formation in the cytochrome oxidase region of the respiratory chain was reconstituted. Proteoliposomes prepared with cytochrome oxidase of bovine heart mitochondria and soybean phospholipids (asolectin) by Hinkle et al.[5, 6, 17] catalysed electron transfer from external ferrocytochrome c to oxygen, the process being greatly stimulated by protonophorous uncouplers, by detergents, or by valinomycin plus nigericin. The latter fact suggested that the oxidation rate in the cytochrome oxidase proteoliposomes is controlled by the membrane potential, a phenomenon similar to respiratory control in intact mitochondria. Further support for this conclusion was obtained when it was observed that respiration in the presence of valinomycin was accompanied by K^+ uptake, which was sensitive to FCCP. K^+ accumulation was coupled with H^+ extrusion; K^+/O and H^+/O ratios were close to 2. Calculation of the magnitude of the membrane potential by Hinkle[17] according to the method of Mitchell and Moyle[8] gave a value of about 90 mV and a transmembrane pH difference of about 1 (alkalinisation inside).

To explain the above data in terms other than involving a membrane potential, one would have to postulate a carrier system catalysing the oxidation-supported electroneutral K^+/H^+ exchange to be operative in the cytochrome oxidase proteoliposomes. However, such an explanation proves to be at variance with the data of experiments in which synthetic penetrating ions, instead of K^+, were used. As was shown by this[23] and Racker's[17] group, tetraphenylphosphonium and tetraphenylarsonium cations can be accumulated, like K^+ ($+$ valinomycin), by the cytochrome oxidase proteoliposomes during oxidation of external cytochrome c (Figure 2.2).

In this laboratory cytochrome oxidase proteoliposomes with cytochrome c inside were also studied[24-26]. External cytochrome c was removed by washing with NaCl. In such preparations, PCB^- accumulated when respiration was initiated by addition of ascorbate and a penetrating hydrogen atom carrier, e.g. phenazine methosulphate (PMS) or tetramethyl-p-phenylenediamine. Subsequent addition of external cytochrome c stimulated the respiration rate

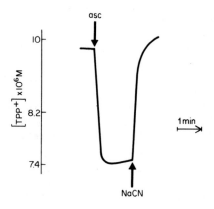

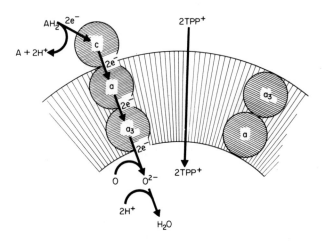

Figure 2.2 Symport of e⁻ and TPP⁺ across the membrane of the cytochrome oxidase proteoliposomes with cytochrome *c* outside. Incubation mixture: 0.25 M sucrose, 0.05 M Tris-HCl (pH = 7.5), 5 mM MgSO₄, 2 × 10⁻⁴ M cytochrome *c*, cytochrome oxidase proteoliposomes (0.5 mg protein ml⁻¹). Additions: 7 mM ascorbate, 1.5 mM NaCN. (From Drachev *et al.*[23], by courtesy of the American Society of Biological Chemists.)

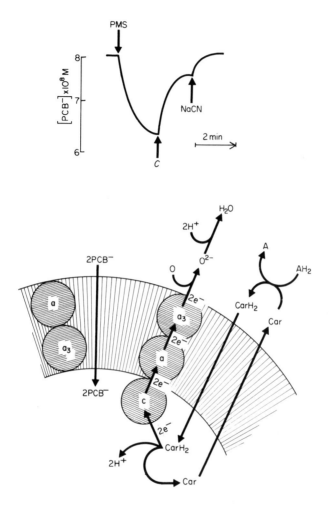

Figure 2.3 Antiport of e⁻ and PCB⁻ across the membrane of
the cytochrome oxidase proteoliposomes with cytochrome *c*
inside. Car, artificial hydrogen atom carrier (PMS or tetra-
methyl-*p*-phenylenediamine). Incubation mixture: 0.25 sucrose,
0.05 M Tris-HCl (pH = 7.5), 5 mM ascorbate, cytochrome
oxidase proteoliposomes containing cytochrome *c* inside (0.4
mg protein ml⁻¹). Additions: 1 × 10⁻⁶ M PMS, 3 × 10⁻⁵ M
cytochrome *c*, 4 mM NaCN. (From Jasaitis *et al.*[24], by courtesy
of Elsevier.)

38

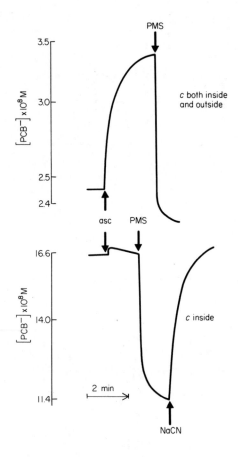

Figure 2.4 Electron transfer-driven PCB^- responses of cytochrome oxidase proteoliposomes containing cytochrome c both inside and outside (upper curve) or inside only (lower curve). Incubation mixture: 0.25 M sucrose, 0.05 M Tris-HCl (pH = 7.5) and cytochrome oxidase proteoliposomes with cytochrome c inside (0.5 mg protein ml^{-1}, upper curve, and 0.6 mg protein ml^{-1}, lower curve). The mixture was supplemented by 2×10^{-4} M cytochrome c (upper curve). Additions: 5 mM ascorbate, 1.3×10^{-6} M PMS, 5.2 mM NaCN. (From Drachev *et al.*[23], by courtesy of the American Society of Biological Chemists.)

and caused the extrusion of accumulated PCB⁻ (Figure 2.3). In proteoliposomes with cytochrome c both inside and outside, ascorbate addition
initiated electron transfer via external cytochrome c and supported PCB⁻
extrusion; the effect was reversed by PMS, which reduced internal cytochrome c (Figure 2.4).

Summarising the above data, one can formulate the following rule: penetrating cations are transferred into, and anions out of, cytochrome oxidase
proteoliposomes if *external* cytochrome c is oxidised; the directions of ion
fluxes are opposite if electron transfer via *internal* cytochrome c occurs.

These results could be predicted by the chemiosmotic hypothesis postulating a transmembranous arrangement of cytochrome oxidase[1]. With this
assumption one can easily explain the above data in terms of a symport of
electrons and penetrating cations (Figure 2.2) or an antiport of electrons
and anions (Figure 2.3).

Direct measurements of the cytochrome oxidase-generated electric potential were carried out by Drachev et al.[23], who studied cytochrome oxidase
proteoliposomes integrated with planar phospholipid membrane. As shown
in Figure 2.5, addition of ascorbate to such a system gives rise to the formation

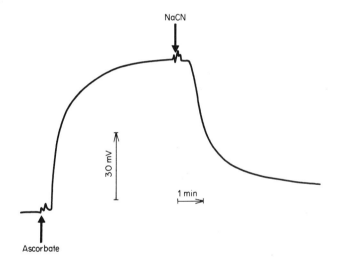

Figure 2.5 Direct measurement of the cytochrome oxidase-
generated electric potential difference. Cytochrome oxidase
proteoliposomes were added into one of two compartments
separated by planar phospholipid membrane made of asolectin.
Electric potential difference across planar membrane was
measured by a voltmeter connected to two Ag|AgCl electrodes
immersed into incubation mixture at both sides of the membrane.
The mixture contained Ca^{2+} ions to induce fusion of proteoliposomes with the planar membrane. Incubation mixture: 0.3 M
sucrose, 5 mM Tris-citrate (pH = 7.2), 30 mM $CaCl_2$, 5 mM
$MgCl_2$, 1×10^{-4} M cytochrome c, and cytochrome oxidase
proteoliposomes (0.3 mg protein ml⁻¹). Additions: 10 mM
ascorbate, 1 mM NaCN. (From Drachev et al.[23], by courtesy of
the American Society of Biological Chemists.)

of an electric potential difference between two aqueous solutions separated by the cytochrome oxidase-containing phospholipid membrane (plus on the ascorbate side of the membrane). Addition of cyanide results in discharge of the membrane potential. Addition of a protonophorous uncoupler decreased the ascorbate oxidation-supported $\Delta\Psi$. The maximal membrane potential values were higher than 100 mV. In experiments with cytochrome oxidase proteoliposomes containing cytochrome c inside, addition of PMS was necessary for the formation of $\Delta\Psi$. In this case, direction of the electric field was opposite to that in proteoliposomes with cytochrome c on the outside.

2.2.2.3 Transmembrane arrangement of cytochrome oxidase

Studies on the localisation of cytochrome oxidase in the mitochondrial membrane confirmed the hypothesis of the transmembraneous arrangement of this enzyme complex[1, 15, 27]. Racker and associates[28] showed that diazobenzene[^{35}S]sulphonate, an impermeant modifier of proteins, labels cytochrome oxidase from both sides of the membrane (intact mitochondria and inside-out submitochondrial particles were compared). Under the same conditions cytochrome c was labelled only in mitochondria, and coupling factor F_1 only in particles. Specific radioactivity of the membranous cytochrome oxidase both in mitochondria and particles was six-fold lower than that of cytochrome oxidase solubilised before diazobenzene[^{35}S]sulphonate treatment, suggesting that the major part (about $\frac{2}{3}$) of the enzyme complex is immersed within the membrane, about $\frac{1}{6}$ on the inner and $\frac{1}{6}$ on the outer surface. The labelling of cytochrome c and F_1 was about the same whether these proteins were attached to the membrane or in solution.

It is firmly established that cytochrome c is attached to the outer surface of the inner mitochondrial membrane. This conclusion is supported not only by experiments with diazobenzenesulphonate[28] but also by many other studies (see Refs. 15, 29). Consequently, the cytochrome c-oxidising component of cytochrome oxidase complex (cytochrome a) should interact with cytochrome c close to the outer surface of the membrane.

There are some indications that the O_2-reducing component of cytochrome oxidase (cytochrome a_3) is localised near the inner membrane surface, i.e. inside mitochondria. Palmieri and Klingenberg[30] showed that conditions favourable for accumulating azide as a weak acid inside mitochondria down the pH gradient are also favourable for the inhibition of respiration. Since the azide-sensitive point of respiratory chain is localised between cytochrome a and O_2, the above result is consistent with a cytochrome a_3 location on the matrix side of the membrane.

An independent line of evidence of the cytochrome oxidase-mediated-transmembrane electron flow is furnished by studies of the degree of cytochrome reduction as the function of the membrane potential in mitochondria. Hinkle and Mitchell[31] demonstrated that a 100 mV membrane potential (negative inside mitochondria) induced a 50 mV negative shift of the mid-point redox potential of cytochrome a as if the latter were located in the middle part of the membrane, namely half-way from cytochrome c to the oxygen-reducing component of the cytochrome oxidase complex. This fact suggests that reverse electron transfer from cytochrome a to c is supported by the transmembranous electric field. It is noteworthy that the effect in question did not depend on the mode of the membrane potential generation: it could be induced by both ATP hydrolysis or downhill movement of K^+ in the presence of valinomycin, or by H^+ in the presence of a protonophorous uncoupler.

This observation was recently confirmed and extended by Wikstrom[32], who presented some evidence that all the steps of the cytochrome c oxidase reaction ($c \rightarrow a$, $a \rightarrow a_3$, and $a_3 \rightarrow O_2$) are affected by energisation of the mitochondrial membrane. One should keep in mind that, according to the chemical energy coupling scheme, $X \sim Y$ formation must be coupled to only one of these three reactions.

The above data, taken together, hardly leave room for doubt as to transmembranous arrangement of the electron transfer process in the cytochrome oxidase region of the respiratory chain. One is therefore bound to assume that the mitochondrial cytochrome oxidase reaction is electrogenic. In fact, electron transfer from ferrocytochrome c to oxygen, being directed across the membrane, should charge the latter until the energy release due to the redox potential difference between cytochromes c and a_3 becomes equal to the energy deficiency due to the movement of electrons against the transmembranous electric field (from positively charged extramitochondrial compartment to the negatively charged matrix space). Hence, the chemical energy of cytochrome c oxidase reaction should be converted into the transmembranous electric potential difference ($\Delta\Psi$). Furthermore, subsequent steps of energy transduction, resulting ultimately in ATP, should be nothing but $\Delta\Psi$ utilisation.

If one wishes to avoid this assumption, the only way out is to postulate the existence of two cytochrome oxidases in the mitochondrial membrane: the first producing the membrane potential and oriented *across* the membrane, and a second, producing an ATP precursor ($X \sim Y$, or energised conformation) and arranged *along* the membrane. To explain the mechanism of the ATP-supported electric potential formation, it could be necessary to propose the involvement of both *trans* and *cis* membranous respiratory chains (such a possibility was considered in detail elsewhere[15]). Although such a complicated concept is not attractive, direct experiments to rule it out would be desirable.

The question whether electron transfer in the coupling sites of respiratory chain is required for the ATP-supported $\Delta\Psi$ formation was answered in experiments with ATPase proteoliposomes, which will be discussed below.

2.2.2.4 Reconstitution of the ATPase generator of membrane potential

In 1971, Kagawa and Racker[2] reconstituted ATPase particles from phospholipids, coupling factors, and hydrophobic proteins required for an oligomycin-sensitive ATP hydrolysis. The particles catalysed a $^{32}P_i$–ATP exchange which was sensitive to both ATPase inhibitors and the agents discharging membrane potential, such as protonophorous uncouplers, valinomycin + nigericin, etc., which stimulated ATPase activity. An ATP-dependent H$^+$ uptake by the particles was also reported[4].

The above mentioned data suggest generation of the membrane potential in the reconstituted particles hydrolysing ATP. We studied particles prepared according to Kagawa and Racker using PCB$^-$ as a probe for $\Delta \Psi$ [24-26a]. The spectrophotometric analysis revealed no cytochrome contaminations in the protein components used for reconstitution. The protein-to-phospholipid ratio of the reconstituted particles was 1:10. Both negative staining

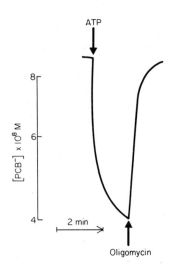

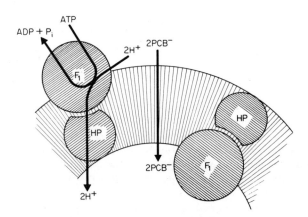

Figure 2.6 ATP-supported PCB$^-$ uptake by ATPase proteoliposomes. F_1, coupling factor F_1; HP, hydrophobic proteins of mitochondrial ATPase. Incubation mixture: 0.25 M sucrose, 0.05 M Tris-HCl (pH = 7.8), 5 mM MgSO$_4$ and ATPase proteoliposomes (1.3 mg protein ml^{-1}). Additions: 1 mM ATP, 6 µg oligomycin ml^{-1}. (From Drachev et al.[26a], by courtesy of the American Society of Biological Chemists.)

and cross-section techniques showed closed vesicles similar to the above described cytochrome oxidase proteoliposomes. The particles catalysed oligomycin- and dicyclohexylcarbodi-imide (DCCD)-sensitive ATP hydrolysis.

ATPase proteoliposomes respond to ATP addition with an uptake of PCB⁻ (Figure 2.6). Oligomycin and uncouplers completely prevented, and reversed if added after ATP, the PCB⁻ uptake, while respiratory chain inhibitors (rotenone, antimycin and cyanide) did not inhibit.

Using the method described above (see pp. 35, 39), Drachev *et al.* showed[26a] that ATP addition to ATPase proteoliposomes attached to a planar membrane in the presence of Ca^{2+} results in the formation of a transmembranous electric potential difference (minus on the ATP side of the membrane). The ATP effect was found to be sensitive to oligomycin (Figure 2.7).

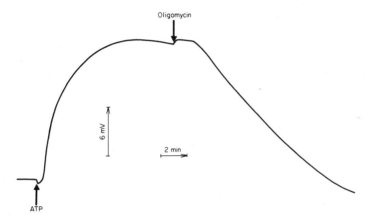

Figure 2.7   Direct measurement of the electric potential difference generated by oligomycin-sensitive ATPase from beef-heart mitochondria. Conditions as in Figure 2.5. ATPase proteoliposomes (0.3 mg protein ml⁻¹) were added instead of cytochrome oxidase ones. Additions: 2.5 mM ATP, oligomycin (30 µg ml⁻¹). (From Drachev *et al.*[26a], by courtesy of the American Society of Biological Chemists.)

These data, together with the results of experiments on cytochrome oxidase proteoliposomes, demonstrate that two types of enzymic generators of membrane potential exist in the mitochondrial membrane: (i) a redox generator exemplified by cytochrome oxidase and (ii) a hydrolytic generator, i.e. ATPase. The first system charges the membrane by transmembranous electron flow. The second seems to be independent of the electron transfer chain. These two systems may be separated and reconstituted with phospholipids to form membranous vesicles with cytochrome oxidase, or, alternatively, the proteins of the oligomycin-sensitive ATPase.

Grinius, in our laboratory[14, 33, 34], showed that the reverse transhydrogenase reaction initiated by addition of NADH and NADP⁺ to the non-energised submitochondrial particles generates a membrane potential of polarity opposite to that induced by forward electron transfer or ATPase activity. To explain this observation in terms of the chemical hypothesis one encounters difficulties connected with the necessity to explain the formation of a membrane potential in spite of a negligible level of any high-energy compounds in the non-energised mitochondrial membrane. Apparently, transhydrogenase reaction could be, like cytochrome oxidase, an example of the $\Delta\Psi$ generation by the redox reaction with no high-energy intermediates involved[29].

On the other hand, redox reactions in the coupling sites of the respiratory chain do not seem to be involved in the ATPase-dependent generation of $\Delta\Psi$. In addition to the data on ATPase proteoliposomes, respiratory chain-deficient submitochondrial particles prepared according to Arion and Racker[35] were shown by Jasaitis et al.[36] to have lost the ability to form $\Delta\Psi$ in the first, third and fourth (transhydrogenase) coupling sites of the redox chain, yet catalyse ATP-supported $\Delta\Psi$ generation. The latter process was resistant to antimycin, and to oxidants or reductants of the respiratory chain.

Evidence for the formation of $\Delta\Psi$ at the cost of ATP energy were obtained in yeast promitochondria by Groot et al.[37]. Promitochondria were found to be deficient in all components of the phosphorylating respiratory chain of mitochondrial inner membrane, i.e. rotenone-sensitive NADH-dehydrogenase, CoQ, non-haeme iron proteins, cytochromes b, c_1, c, a and a_3. Therefore, any effect of ATP in this system must be independent of the operation of the respiratory chain.

We conclude from the above data that $\Delta\Psi$ can be formed by the respiratory chain, or, alternatively, by the ATPase system. As was found by Mitchell's and our group[8-11, 34], both the direction and magnitude of $\Delta\Psi$ generated by these two systems in mitochondria are the same. The total values for the electrochemical H^+ potential of the membrane of respiring mitochondria and submitochondrial particles are of the same order of magnitude, and energetically similar to that of the ATP-driven reaction. This means that the ATPase system, if reversible, could form ATP at the cost of $\Delta\mu_H$ produced by respiration. This possibility was recently demonstrated by experiments on reconstitution of proteoliposomes competent in oxidative phosphorylation.

2.2.2.5   *Reconstitution of proteoliposomes catalysing oxidative phosphorylation*

Racker and Kandrach[3] were the first to succeed in reconstituting oxidative phosphorylation from solubilised enzyme proteins and phospholipids. To this end, proteoliposomes with both cytochrome oxidase (+ cytochrome c inside) and ATPase were prepared. The reconstituted particles catalysed ATP formation coupled to the oxidation of ascorbate + PMS. Phosphorylation was inhibited by cyanide or oligomycin, as well as by protonophorous uncouplers, valinomycin + nigericin, and by external cytochrome c.

The study of this system by our group[23, 26] using the PBC$^-$ method gave the following results. Cytochrome oxidase and ATPase proteoliposomes were shown to be competent in the membrane potential of a formation (plus inside) depending on respiration or ATP, respectively (Figure 2.8). $\Delta\Psi$ generation supported by oxidation of ascorbate required addition of PMS to reduce internal cytochrome c. Treatment with external cytochrome c in the presence of ascorbate decreased $\Delta\Psi$ in samples with ATP as well as with PMS.

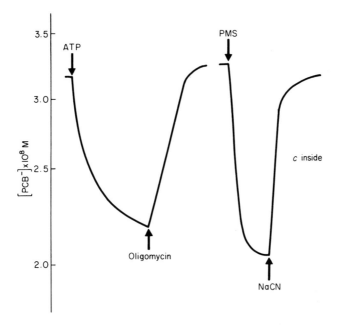

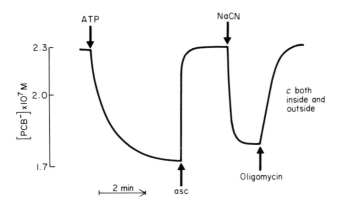

Figure 2.8 PCB⁻ responses of ATP + cytochrome oxidase proteoliposomes. Incubation mixture: 0.25 M sucrose, 0.05 M Tris-HCl (pH = 7.5), ATPase + cytochrome oxidase proteoliposomes (0.9 mg protein ml⁻¹). Lower curve: mixture was supplemented with 2×10^{-4} M cytochrome c. Additions: 1 mM ATP, 1.6 μg oligomycin ml⁻¹, 4×10^{-6} M PMS, 1 mM NaCN, 5 mM ascorbate. (From Drachev et al.[26a], by courtesy of the American Society of Biological Chemists.)

Addition of cytochrome c plus ascorbate prior to PMS resulted in formation of $\Delta\Psi$ of the opposite sign (plus outside). Subsequent PMS treatment reversed the effect of external cytochrome c. In all cases, regardless of the generation mechanism and direction of electric field, $\Delta\Psi$ decreased when a protonophorous uncoupler was added.

Summarising the facts presented in this section, one can conclude that the system of oxidative phosphorylation can be reconstituted from two $\Delta\Psi$ generators (redox and hydrolytic) incorporated into the membrane of phospholipid vesicles. $\Delta\Psi$ of the proper direction formed on membranes of these vesicles proves to be a necessary intermediate between oxidation and phosphorylation. The latter conclusion is supported by the following facts. (i) Any agents discharging the proton motive force, including protonophorous uncouplers, detergents and such specific ion transport systems as valinomycin $+$ nigericin $+$ K^+, inhibit oxidative phosphorylation in both intact mitochondria and reconstituted proteoliposomes. (ii) In proteoliposomes, enzymatic generation of the membrane potential of opposite polarity inhibits phosphorylation coupled with ascorbate oxidation via internal cytochrome c. This effect can be obtained by addition of external cytochrome c. It should be stressed that this is not an injuring side action of cytochrome c on the proteoliposome membrane, since, under the same conditions, external cytochrome c induced an oxidation-dependent K^+ and TPP^+ influx into the vesicles, as well as PCB^- extrusion from the proteoliposomes.

An important consequence of the above conclusions is that the role of the redox chain in ATP formation, being restricted to furnishing a membrane potential, is rather non-specific. Hence, the cytochrome oxidase generator in a reconstituted system of oxidative phosphorylation can be substituted by any other mechanism of membrane potential generation. This point was supported by the observation of Ragan and Racker[7], who reconstituted oxidative phosphorylation in the first coupling site with NADH dehydrogenase and oligomycin-sensitive ATPase. The P:2e^- ratio of NADH oxidation by CoQ_1 in the NADH dehydrogenase plus ATPase proteoliposomes was 0.5.

2.2.2.6   Bacteriorhodopsin H^+ pump

The non-specific role of respiration in the formation of an electrochemical H^+ potential required for ATP synthesis was recently demonstrated in experiments with a bacteriorhodopsin proton pump.

Stoeckenius and associates[38-41] studied membranes of the extreme halophilic *Halobacterium halobium*. These bacteria contain a protein closely resembling rhodopsin, the visual chromoprotein of higher animals. This protein, called bacteriorhodopsin, is composed of an opsin-like polypeptide of 26 000 molecular weight which forms a Schiff base with retinal. Hypotonic disruption of the bacteria yielded a mixture of membrane fragments, from which a fraction of purple round or oval sheets ~0.5 µm in diameter and ~50 Å thick was isolated. These purple membranes were composed of bacteriorhodopsin as a single protein component (75% of dry weight) and phosphoglycerol ether of dihydrophytol (25%). A low-angle x-ray diffraction study showed a hexagonal lattice, the distance between the centres of the adjacent hexagons being 63 Å. Illumination of the purple membranes induced a shift of the bacteriorhodopsin absorption spectrum characteristic of the light response of animal rhodopsin and an H^+ release. The synthesis of bacteriorhodopsin in halobacteria drastically increased at the end of the logarithmic growth phase when the oxygen level in the medium was low and the bacterial cell concentration high. Under these conditions, turning on of light causes an inhibition of oxygen consumption, an H^+ efflux and an increase in the intracellular ATP level. All light effect in intact cells (but not in the bacteriorhodopsin sheet suspension) were inhibited by protonophorous uncouplers.

The following experiments carried out by Racker and Stoeckenius[42] showed that bacteriorhodopsin-containing proteoliposomes can be reconstituted from purple sheets and soybean phospholipids. On illumination,

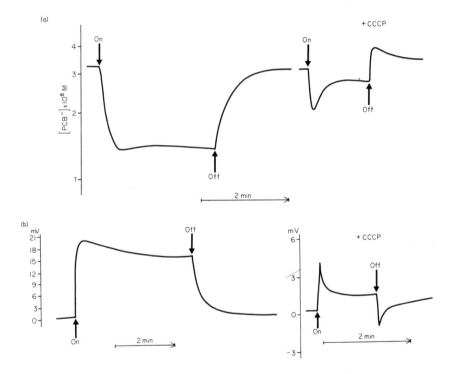

Figure 2.9 PCB$^-$ responses of bacteriorhodopsin proteoliposomes (a) and direct measurement of the electrogenesis in the planar phospholipid membrane with bacteriorhodopsin inclusion (b). For (a) the incubation mixture contained: 0.3 M sucrose, 5 mM Tris-citrate (pH = 6.25), bacteriorhodopsin proteoliposomes (0.08 mg protein ml^{-1}). For (b) the planar artificial membrane was formed from the mixture containing decane solution of azolectin and bacteriorhodopsin sheets. E.m.f. was measured by Ag|AgCl$_2$ electrodes connected with vibron electrometer; electrodes were immersed into two membrane-separated compartments containing 0.15 KCl. CCCP concentration (right curves) was 2×10^{-7} M. (From Kayushin and Skulachev[43], and from Drachev et al.[43a], by courtesy of North Holland Publishing Co.)

proteoliposomes took up H^+ ions which were released in the dark. Valinomycin accelerated, whereas a protonophorous uncoupler inhibited these pH
responses. Addition of the oligomycin-sensitive ATPase proteins from
bovine heart mitochondria during reconstitution yielded proteoliposomes
which catalysed photophosphorylation of ADP by inorganic phosphate.
The authors concluded that bacteriorhodopsin–ATPase proteoliposomes
'represent a simple model system for a biological proton pump capable of
generating ATP from ADP and P_i.'

Experiments of our group and of Kayushin's laboratory in the Institute
of Biophysics in Pustchino revealed the electrogenic character of the bacteriorhodopsin proton pump. It was found[43] that bacteriorhodopsin proteoliposomes take up PCB^- in a light-dependent fashion (Figure 2.9a). The effect
was strongly inhibited by low (uncoupling) amounts of the protonophore
trichlorocarbonyl cyanide phenylhydrazone (CCCP).

Parallel measurements of atebrin fluorescence, which is a probe[18] for transmembrane ΔpH, showed the light-induced fluorescence decrease, indicating
acidification of the intraproteoliposomal compartment. The pH level of the
incubation medium, in agreement with the data of Racker and Stoeckenius[42],
was found to move to the alkaline side.

These results are sufficient to conclude that bacteriorhodopsin is operative
as an electrogenic proton pump generating $\Delta\mu_H$ composed of $\Delta\Psi$ and ΔpH
(plus and acid inside the proteoliposomes).

Generation of the photoelectromotive force by bacteriorhodopsin has
been directly demonstrated by our group in experiments with a planar
phospholipid membrane[43a]. Bacteriorhodopsin sheets can be incorporated
into such a membrane by mixing them with a decane solution of asolectin,
which forms an artificial membrane on the aperture in the wall of a Teflon
vessel. Measurements of the electric parameters of this membrane by standard
electrometric techniques revealed no electric potential difference in the dark.

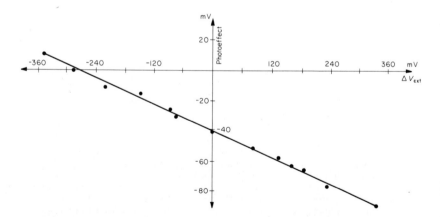

Figure 2.10 Bacteriorhodopsin-mediated photoeffect as the function of the external
ΔV. Incubation mixture: 0.3 M sucrose, 0.09 M $(NH_4)_2SO_4$, 5 mM Tris-citrate:
bacteriorhodopsin proteoliposomes (0.1 mg protein ml^{-1}). (From Drachev et al.[43b],
by courtesy of the American Society of Biological Chemists.)

Light induced generation of a membrane potential (Figure 2.9b), which was strongly inhibited by 2×10^{-7} mol l^{-1} CCCP.

Studying bacteriorhodopsin proteoliposomes associated with a planar membrane, Drachev and Semenov observed photoeffects as high as 150 mV. It was affected by an electric potential difference generated across the planar membrane by an external source (ΔV_{ext}). A ΔV_{ext} of the same sign as that of the photo-e.m.f. decreased, but one of the opposite sign increased, the light-induced response (Figure 2.10). At a certain ΔV_{ext} the photoeffect was abolished. At this point, ΔV_{ext} should be equal to the photo-e.m.f. if all bacteriorhodopsin generators have the same orientation. The photo-e.m.f. should be higher than this value if some of the generators are oriented in the opposite direction. According to Figure 2.10, the photo-e.m.f. is >300 mV.

The fact that bacteriorhodopsin, a rather small protein composed of one polypeptide chain, carries out a proton-translocating function in a membrane where there are no other proteins renders it highly improbable that an electron transfer chain is required for *H. halobium* photophosphorylation. Thus, ATP formation by bacteriorhodopsin–ATPase proteoliposomes should be the result of $\Delta\mu_H$ formation and utilisation with no redox chain involved.

2.2.2.7 Ion transfer phosphorylation

All active ion transport systems thus far studied proved to be capable of ATP formation coupled with reverse (downhill) ion movement. Garrahan and Glynn[44] described ATP synthesis in erythrocytes driven by a downhill Na^+ influx and K^+ efflux, i.e. Na^+/K^+ antiport of the direction opposite to that coupled with the Na^+, K^+–ATPase-catalysed ATP hydrolysis. Jagendorf and Uribe[45] reported formation of ATP coupled to an H^+ flux down a pH gradient in chloroplasts in the dark. A year later, Cockrell *et al.*[46] showed ADP phosphorylation in mitochondria which was coupled with the downhill valinomycin-mediated K^+ efflux. Barlogie *et al.*[47, 48] and Panet and Selinger[49] demonstrated ATP formation with a Ca^{2+} efflux from Ca^{2+}-loaded vesicles of sarcoplasmic reticulum.

Since these pioneering studies were published, observations on the ion transfer-driven phosphorylation have been confirmed and extended in several laboratories (for reviews, see Refs. 15 and 29). Results of studies along this line allow one to regard reversibility of energy transduction to be inherent in active ion transport systems of biomembranes.

2.2.3 The current state of the chemiosmotic hypothesis

The data summarised in the above sections are sufficient, in our opinion, for stating that chemiosmotic principle of energy coupling has been experimentally proven at least for the cytochrome oxidase reaction. Advocates of the chemical (conformational, etc.) hypothesis have to resort to a very complicated scheme, according to which alternative mechanisms of oxidative phosphorylation, chemiosmotic and, e.g. chemical, coexist in the same mitochondrial membrane, equation (2.3):

$$ATP \leftarrow oxidation \rightarrow \Delta\mu_H \rightarrow ATP \qquad (2.3)$$

Such a concept, like any compromise, might explain some observations which still remain unexplained in terms of the version of the chemiosmotic hypothesis as presented by Mitchell in 1966. However, one should keep in mind that no facts have been described inconsistent with the *main principle* of the chemiosmotic scheme, and that all the systems of the redox chain phosphorylation described in the literature required closed vesicular membranous particles. As to publications on electron transfer phosphorylation in soluble enzymic systems appearing from time to time during the past decade, they invariably have turned out to be erroneous. Up to date, no reproducible valinomycin plus nigericin-resistant enzymic system of electron transfer phosphorylation has been reported. Membraneous vesicles of low permeability for H^+ seem to be absolutely necessary for respiratory and photosynthetic ATP formation.

Some authors reported values of stoichiometry and energetics of ion transfer and oxidative phosphorylation differing from those predicted by Mitchell (for review, see Refs. 15, 29 and 50). However, Hinkle *et al.*[5,17,51], having reinvestigated this problem, clearly showed that, in such simple systems as submitochondrial particles or reconstituted proteoliposomes, the above parameters harmonise well with the theoretical ones. The discrepancies between the theoretical predictions and the data of some experiments with mitochondria and chloroplasts were probably the result of working with more complicated systems. There are some indications that oxidative enzyme complexes interact with phosphorylating enzymes in coupling membranes. For instance, the interaction of antimycin A with the cytochrome *b* region of mitochondrial respiratory chain was found to change the affinity of coupling factor F_1 to aurovertin, and, *vice versa*, addition of aurovertin affects the antimycin binding[52]. There is no simple explanation for these observations in terms of the chemiosmotic hypothesis. However, these findings are probably a consequence of the structural organisation of the membrane rather than of the energy coupling mechanism *per se*. An inhibitor-induced conformation change in one membraneous enzyme protein may affect the property of another protein localised in its vicinity. We should keep in mind that membraneous proteins are not only catalysts but also building materials composing the major part of the mitochondrial membrane. Negative staining and aurovertin titrations show that the F_1 content per unit of the membrane area is so high that the ATPase complex is likely to have contact with some neighbouring membraneous proteins. Such protein–protein interactions might result in some change in the properties of a given enzyme when a neighbouring enzyme is modified by an inhibitor.

Particular features of Mitchell's original hypothesis, e.g. versions of the $\Delta\mu_H$ formation by electron transfer and ATP hydrolysis, have been criticised. It would be naive to think that all details of such a complex multi-component mechanism as the membraneous energy coupling system can be predicted by a basic hypothesis. The advantages of the chemiosmotic hypothesis are so great that, we believe, we are justified in accepting this concept as a working hypothesis and to analyse specific problems of oxidative phosphorylations. The problems in question are: (i) how oxidative enzymes form $\Delta\mu_H$, and (ii) how ATP-synthetase utilises $\Delta\mu_H$. Perspectives of investigations along these lines will be considered below.

2.3   THE MECHANISMS OF FORMATION AND UTILISATION OF THE ELECTROCHEMICAL POTENTIAL OF H$^+$

2.3.1   The mechanism of membrane charging by cytochrome oxidase

Among different membrane potential generators, cytochrome oxidase seems to have been elucidated to a level allowing the main steps of the energy transduction process to be rationalised.

As noted above, the membrane charging by cytochrome oxidase is most likely a result of the transmembraneous electron movement. To catalyse such a process, cytochrome oxidase proteins must be incorporated into the central hydrophobic layer of the membrane responsible for the high electric resistance of this system. To avoid a decrease in the resistance, these proteins must be hydrophobic themselves and carry out the electron transfer in a manner allowing the membrane structure to retain its insulating properties.

The latter result may be achieved in different ways. One of the possibilities is the following. Cytochrome oxidase proteins are firmly fixed in the membrane. No conformation changes, rotations, etc., are permitted. In this case, it is rather easy to imagine how the phospholipid bilayer of the membrane can be interrupted by protein (cytochrome oxidase) components without a marked decrease in the resistance. However, it is not clear what type of electron transfer mechanism might be operative in a system organised in such a fashion. It seems to be improbable that the mechanism in question is related to semiconductor phenomena, although they have been discussed by some authors with reference to cytochrome oxidase[53]. Significant conformational changes were found to accompany the transition of cytochrome oxidase from the oxidised to the reduced state[54]. Rotation of cytochrome a_3 in the mitochondrial membrane was also reported[55]. Taking into account these observations, we may postulate that conformational changes of haeme-containing polypeptides of cytochrome oxidase are involved in the electrogenic oxidoreduction. Some of the protein components which are present in the cytochrome oxidase complex may be firmly fixed in the membrane, forming a cage which may allow conformational changes of haemoproteins without deterioration of the insulating properties of the hydrophobic membraneous phase. These cage-forming ('structural') proteins should be extremely hydrophobic to permit the intimate contact between phospholipid bilayer and cytochrome oxidase. A very stable complex of hydrophobic proteins with phospholipids may be identical to the proteolipid found in cytochrome oxidase[56]. One may think that these 'cage proteins' are synthesised in the mitochondria, whereas less hydrophobic haeme-containing polypeptides are made in the cytoplasm (see Refs. 57 and 58).

Figure 2.11 shows a tentative scheme of the cytochrome oxidase complex in the membrane. Vertical shading indicates hydrophobic parts of proteins;

horizontal shading, those of phospholipids; and two-directional shading, proteolipids. It is proposed that reduction of cytochrome a by cytochrome c induces a conformational change designated tentatively by rotation of the cytochrome a globule. As a result, electrons are transferred from the outer surface of the mitochondrial membrane to its middle part. Further electron transfers from cytochrome a to a_3 are postulated to induce a conformational change in cytochrome a_3, resulting in an electron being carried to the opposite (inner) surface of the membrane where the reduction of oxygen takes place.

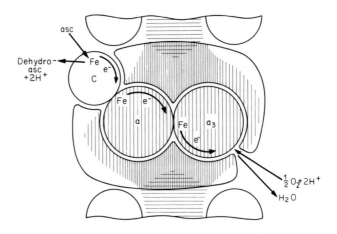

Figure 2.11   A possible arrangement of the cytochrome oxidase in membrane. Vertical shading, hydrophobic parts of proteins; horizontal shading, those of phospholipids; two-directional shading, proteolipids

The question arises what is the mechanism of the oxidation (reduction)-induced conformational changes of the cytochrome oxidase subunits. As a working hypothesis, the scheme shown in Figure 2.12 may be used. In this scheme it is postulated that reduction of haeme a_3 by a (stage 1) entails the appearance of a negatively charged anionic group A_3^- on the cytochrome a_3 protein localised close to another anionic group, A_2^-, on cytochrome a protein. Thus, exergonic electron transfer from cytochrome a to a_3 should be accompanied by formation of a local electric field in the $A_2^- - A_3^-$ region, so that the free energy released by oxidoreduction is converted to the potential energy of the electric repulsion of A_2^- and A_3^-. The next stage (2) is a conformational change of cytochrome a_3 (tentatively, rotation of cytochrome a_3 protein globule) resulting in A_3^- being removed from A_2^- and the negative charge (A_3^-) travelling a half-membrane distance. The A_3^- movement is directed to the side of the mitochondrial interior, which charges negatively. At this stage, the energy of the local electric field is transduced into the energy of the transmembrane electric field. The reaction sequence is completed by the transfer of electron from haeme a_3 to oxygen (stage 3) and rotation of cytochrome a_3 (stage 4). It is noteworthy that, at stage 4, anionic group A_3^- returns to the initial position in the neutralised form.

The scheme of the cytochrome $a-a_3$ interaction given in Figure 2.12 represents only one of the possible versions of this mechanism. Perhaps forces other than electrostatic repulsion induce a conformational change of cytochrome a_3. The main idea of the scheme illustrated by Figure 2.12 is that

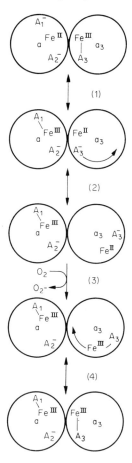

Figure 2.12 A tentative scheme of the mechanism of electron transfer from cytochrome a^{2+} to oxygen

electron transfer from cytochrome a to a_3 down the redox potential gradient results in some energetically unfavourable state of the cytochrome $a-a_3$ complex which then changes to a favourable one, the latter process being accompanied by a charge transfer against a transmembranous electrical gradient.

2.3.2 Reversible H$^+$–ATPase of coupling membranes

2.3.2.1 Definition of reversible H$^+$–ATPase

Transmembranous electron flow catalysed by the cytochrome c–cytochrome oxidase complex should result in an electric potential difference and a pH

difference across the membrane. The latter effect is due to the nature of the
cytochrome oxidase reductant and oxidant which are a hydrogen donor
(ascorbate in the above experiments) and the hydrogen acceptor (O_2),
respectively.

Oxidation of any hydrogen donor by cytochrome c–cytochrome oxidase
complex is a charge-separating process: the electron of the H atom of the
donor is transferred to cytochrome c, while the proton is released into the
solution. The latter effect results in the cytochrome c (extramitochondrial)
space being acidified. On the other hand, electron transfer from cytochrome
oxidase to O_2 leads to alkalinisation of the cytochrome a_3 compartment
(matrix space), which is accompanied by addition of H^+ ions of the intra-
mitochondrial solution to the ionised oxygen molecule. If the membrane
of mitochondria is impermeable for the ions of the medium, the cytochrome
oxidase-released energy is mainly converted into a transmembranous electric
potential difference ($\Delta\Psi$). The pH changes involved are limited by the amount
of electrons which are to be transferred across the membrane to charge its
capacity. If there are some penetrating ions in the medium, ΔpH becomes the
major component of the cytochrome oxidase-produced electrochemical
potential of H^+. Therefore, mitochondrial ATP–synthetase should be
organised as the system whose operation can be supported by both $\Delta\Psi$ and
ΔpH. This would be the case if ATP formation were coupled with hydrogen
ion movement down the total electrochemical H^+ gradient ($\Delta\mu_H$).

According to Mitchell[1], the ATP synthesis coupled with the redox chain
electron transport represents a simple reversal of a $\Delta\mu_H$-producing ATPase
reaction. We would like to define the enzyme complex catalysing this reaction
as H^+–ATPase (or ATPase, carrying out active transport of H^+ ions) by
analogy with other transport ATPases, such as Na^+,K^+–ATPase or Ca^{2+}–
ATPase.

Several lines of evidence suggest that the ATP-supported charging of the
mitochondrial membrane is due to H^+ transport against $\Delta\mu_H$. It was demonstrated that, in the presence of any penetrating ion, ATP hydrolysis always
results in a ΔpH arising across the mitochondrial membrane. This effect
takes place regardless of the kind of ions or ionophores used: both natural
and synthetic compounds can be applied (for review, see Refs. 15 and 29).
This property of the ATPase system makes it improbable that the ΔpH
formation is a result of, e.g. a cation–H^+ exchange. In other words, the ΔpH
seems to be caused by an active transport of H^+, rather than by the coopera-
tion of a cation (anion ?)-transporting ATPase and a nigericin-like ionophore.
Otherwise, one has to assume that both the ATPase and the ionophore can
transport synthetic ions of different structures. It is noteworthy that the
ATP-dependent formation of ΔpH also occurs in simple reconstituted
ATPase proteoliposomes.

2.3.2.2   The membrane potential as a 'non-phosphorylated high-energy intermediate' of the H^+–ATPase reaction

In the old schemes of 'energy transfer chain', electron transfer and ATP
synthesis were connected via a number of high-energy intermediates ($I\sim X$,

$X \sim Y$, etc.). The main argument in favour of the existence of these intermediates was the fact that respiratory chain energy can be effectively utilised for such energy-consuming functions as reverse electron transport or ion accumulation with neither inorganic phosphate nor adenine nucleotide being involved. It was assumed that the mentioned functions are supported by $X \sim Y$ hydrolysis which, unlike ATP formation, is insensitive to oligomycin:

$$\text{Oxidation} \leftrightarrow X \sim Y \xleftarrow{\quad \text{oligomycin} \quad} \text{ATP}$$
$$\updownarrow$$
$$\text{Ion accumulation,} \qquad\qquad (2.4)$$
$$\text{reverse electron}$$
$$\text{transport}$$

Assuming that $\Delta\mu_H$ is an intermediate, coupling oxidation and phosphorylation, one can modify equation (2.4) as follows:

$$\text{Oxidation} \leftrightarrow \Delta\mu_H \xleftarrow{\quad \text{oligomycin} \quad} \text{ATP}$$
$$\updownarrow$$
$$\text{Ion accumulation,} \qquad\qquad (2.5)$$
$$\text{reverse electron}$$
$$\text{transport}$$

According to equation (2.5) the role of the non-phosphorylated high-energy intermediate is performed by $\Delta\mu_H$.

It is still not excluded that some high-energy intermediates may participate in the energy transduction between $\Delta\mu_H$ and ATP. Such a scheme was proposed by Mitchell[1]. However, there are no experimental reasons to complicate the hypothesis in this way. Recently, Mitchell[59] and, independently, Glagolev[60] in our group put forward simpler schemes requiring no $X \sim Y$. In terms of this concept, ATP hydrolysis by F_1 is coupled with the H^+-translocation catalysed by the oligomycin-sensitive ATPase. High-energy compounds are not necessarily involved in ATP hydrolysis (or synthesis) if F_1 action gives rise directly to the charging of the membrane.

Indeed, recent results of Ryrie and Jagendorf[61] suggest that CF_1 (chloroplast factor F_1) is intimately involved in $\Delta\mu_H$ utilisation (formation). The authors found that energisation of the chloroplast membrane, i.e. generation of $\Delta\mu_H$, initiates CF_1 conformational changes which can be revealed by measuring tritium incorporation into CF_1. The control of CF_1 from chloroplasts, incubated with tritium without light, contained 5 tritium ions per CF_1 molecule. In the light or following acid–base treatment, the tritium ion: CF_1 ratio was about 90. Addition of ADP + P_i decreased this ratio to 50. The effects of both light and ADP + P_i were insensitive to Dio-9 and phlorizin, inhibitors which influence the same step of energy transduction in chloroplasts as oligomycin in mitochondria, i.e. $\Delta\mu_H \leftrightarrow$ ATP. This important result indicates that conformation of CF_1 and, hence, its functioning is affected by $\Delta\mu_H$ in a Dio-9 resistant fashion. Further study is necessary to elucidate where the above effect is due directly to the action of $\Delta\mu_H$ on F_1 or whether the $\Delta\mu_H$ influence is mediated by changes of neighbouring components of the chloroplast membrane.

2.3.2.3   Resolution of the H^+–ATPase complex into catalytic (ATPase) and H^+–translocating components

The proteoliposome experiments made it possible to separate and demonstrate independently the H^+-translocating and ATP-hydrolysing activities of the mitochondrial ATPase complex. It was found that incorporation of hydrophobic proteins of the oligomycin-sensitive ATPase into proteoliposomes greatly increased the H^+ conductivity. Addition of oligomycin to such proteoliposomes decreased the H^+ conductivity. Apparently, hydrophobic proteins contain an oligomycin-sensitive proton-translocating component which is required for the ATPase to be coupled with $\Delta\mu_H$ generation. The ATPase, *per se*, can be detached from the hydrophobic proteins in the form of coupling factor F_1 which splits ATP in an oligomycin-resistant fashion. In the earlier experiments with F_1-deficient submitochondrial particles, it was observed that oligomycin improves energy-linked functions such as respiration-driven reverse electron transfer (for review see Refs. 29 and 62). PCB$^-$ experiments showed that in these particles oligomycin prevented dissipation of $\Delta\Psi$ formed by any coupling site of respiratory chain[10, 29, 34]. The 'coupling' effect of oligomycin is accompanied by a decrease in the H^+ conductivity of the particles' membrane, as was discovered by Mitchell *et al.*[63] (see also Ref. 64). The H^+ conductivity induced by added protonophorous uncouplers was not affected by oligomycin[29]. The insulating properties of the membrane of F_1-deficient particles can also be improved[29, 34] by reconstitution of the particles with F_1.

Summarising the results noted above, we can conclude the following. (1) Hydrophobic proteins dissociated from mitochondrial ATPase facilitate the downhill transmembrane H^+ movement in an oligomycin-sensitive manner. (2) ATPase dissociated from hydrophobic proteins (coupling factor F_1) hydrolyses ATP in an oligomycin-insensitive manner. (3) Reconstitution of H^+-translocating hydrophobic proteins and F_1 yields a system which catalyses an oligomycin-sensitive ATP $\leftrightarrow \Delta\mu_H$ energy transduction.

2.3.2.4   A tentative scheme of H^+–ATPase

In one of the previous sections (see p. 53), the mechanism of $\Delta\mu_H$ formation by cytochrome oxidase was considered. The chemistry of the ATPase reaction clearly differs from that of cytochrome oxidase. Therefore, the mechanisms for $\Delta\mu_H$ generation by these two systems should be different, at least in some respects. It does not mean, however, that the difference inevitably affects the fundamental principle of the membrane charging process. It seems reasonable to consider the H^+–ATPase in terms of the concept applied above to the cytochrome oxidase generator. The tentative scheme of this mechanism is given in Figure 2.13. It is proposed that the process is carried out by a two-part system composed of a catalytic (ATPase) subunit and a H^+-translocating subunit. The latter resembles the hypothetical rotating globules of cytochrome oxidase haemoproteins. The reaction can be initiated by adding two H^+ ions to the hydrophobic globule (stage 1)˙ and one ATP molecule to the ATPase component (stage 2). It is suggested that ATP hydrolysis (stage 3) causes the formation of a local electric field and repulsion between one of the negatively charged reaction products (e.g. ADP—O$^-$), bound to the ATPase subunit, and an anionic group of the other subunit (A$^-$). Rotation (or some other conformational change) of the H$^+$-translocating globule with hydrogen ions attached to it (stage 4) results in the

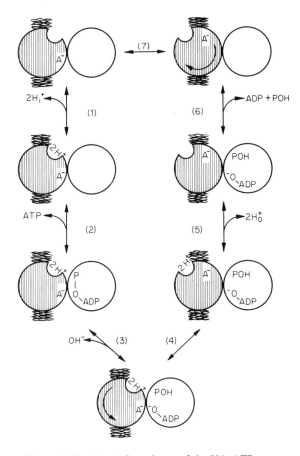

Figure 2.13   A tentative scheme of the H$^+$–ATPase

removal of A$^-$ from ADP—O$^-$ and translocation of 2H$^+$ across the hydrophobic barrier of the membrane. Then 2H$^+$ are released on the opposite side of the membrane (stage 5), ADP and P$_i$ dissociate from the ATPase subunits (stage 6) and the hydrophobic globule returns to the initial position (stage 7). The same reaction sequence, in reverse, can catalyse ATP synthesis coupled with the downhill H$^+$ movement.

One may suggest that the catalytic component of this system is F$_1$, and the H$^+$-translocating globule is one of hydrophobic proteins of H$^+$–translocase, e.g. the proteolipid which combines with dicyclohexylcarbodi-imide[65, 66].

In the native complex, H^+–ATPase activity of H^+-translocating component should be controlled by the catalytic component to prevent the downhill diffusion of H^+ without ATP formation. When F_1 is detached, we propose that this control disappears so that H^+-translocation is now independent of phosphorylation. This leads to the discharge of the membrane potential produced by the respiratory chain. Oligomycin treatment inhibits the H^+–translocase activity and allows the respiration-produced membrane potential to be maintained in spite of the lack of F_1. It is noteworthy that such an effect of oligomycin can be demonstrated under conditions that preclude the formation of high-energy compounds. In our group, Severina[34, 67] showed an oligomycin-induced increase in the membrane potential during transhydrogenase action when the energy yield was less than 1 kcal mol^{-1} of oxidised NADPH. This observation is inconsistent with the concept of the 'coupling' action of oligomycin by inhibition of the hydrolysis of a high-energy intermediate[62, 63].

To conclude this section, we would like to note that the postulated scheme of H^+–ATPase may be applied to ATPases catalysing active transport of ions other than H^+. To this end, it would be sufficient to suggest that the ion-translocating globule binds these ions, instead of H^+. The Ca^{2+}–ATPase of sarcoplasmic reticulum should bind Ca^{2+}, and the Na^+,K^+–ATPase of the plasma membrane of animal cells should bind Na^+ and K^+ (in the latter case, the Na^+- and K^+-binding sites should be on the opposite sites of the globule). It is noteworthy that Na^+,K^+–ATPase, like H^+–ATPase, is sensitive to such a specific inhibitor as oligomycin[68]. Another property common for H^+–, Na^+,K^+–ATPase and Ca^{2+}–ATPase is reversibility of the ATPase system.

2.3.3   Transhydrogenase generator of membrane potential

Formation of membrane potential by transhydrogenase reaction was shown in submitochondrial particles and bacterial chromatophores by means of the PCB^- probe[10, 12, 33, 69]. Conventionally, this system may be classified, like cytochrome oxidase, as a generator of the redox type. Nevertheless, the transhydrogenase mechanism clearly differs from that of cytochrome oxidase.

(a) In the transhydrogenase reaction, the reducing equivalents which are transferred from the donor (NADPH) to the acceptor (NAD^+) is H^-, instead of e^- transferred by cytochrome oxidase. (b) Both the donor and acceptor interact with transhydrogenase on the same (outer in the case of submitochondrial particles) side of the membrane when the membrane potential is generated[10, 70]. (c) The H^- transfer between NADP and NAD catalysed by mitochondrial[71, 72] (as well as chromatophore[73]) transhydrogenase occurs without the exchange of the protons, which participate in the oxidoreduction, with those of water.

The latter fact indicates that the transhydrogenase oxidoreduction *per se*, even if directed across the membrane, cannot result in the formation of $\Delta\mu_H$. Therefore, a mechanism of membrane charging other than that used in the third energy coupling site should be considered.

We have discussed above the mechanism of $\Delta\mu_H$ generation by H^+–ATPase. In this system, as in the transhydrogenase, both the substrate (ATP) and the products (ADP and P_i) interact with the enzyme on the same side of the membrane, and the $\Delta\mu_H$ production seems to be a result of some additional process (H^+ translocation), coupled with the exergonic chemical

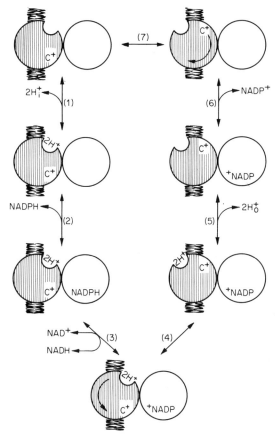

Figure 2.14 A possible mechanism of the membrane potential formation in the fourth (transhydrogenase) energy coupling site of respiratory chain

reaction, rather than of the reaction itself. This suggests that H^+–ATPase and transhydrogenase generators may be organised in a similar way.

A scheme considering transhydrogenase in these terms is shown in Figure 2.14. We suggest that transhydrogenase is composed of a catalytic and an H^+-translocating subunit, the latter bearing a positive charge C^+ near the NADPH-binding site of the former. Oxidation of the bound NADPH by NAD^+ giving rise to bound $NADP^+$ might induce rotation of the H^+–translocase globule and the charging of the membrane by the mechanism proposed above for H^+–ATPase.

2.3.4   Membrane potential generation in the first and the second energy coupling sites

In the scheme, postulated originally by Mitchell[1], the first and second coupling sites of respiratory chain were suggested to be organised like cytochrome oxidase, generating the membrane potential by means of electron transfer across the membrane. Accepting this point of view it must be assumed that the flow of reducing equivalents from NADH to O_2 traverses the membrane six times; thereby electron transfer reactions alternate with hydrogen atom transfer reactions. Three energy coupling sites were postulated to correspond to three loops of the respiratory chain.

In our laboratory, an attempt was made to verify the loop hypothesis[70]. The membrane potential was measured using the PCB^- method. A non-penetrating electron donor (NADH) and acceptor (ferricyanide) were tested in cyanide- and antimycin-inhibited mitochondria and inside-out submitochondrial particles. In the latter system, NADH was rapidly oxidised by ferricyanide. The rotenone-sensitive portion of the oxidation was coupled with formation of a membrane potential. In mitochondria, added NADH is not oxidised. Endogenous NADH was oxidised by ferricyanide only if antimycin was omitted. Again, oxidoreduction between NADH and ferricyanide was competent in generating a membrane potential. Under the same conditions on substitution of ferricyanide by a penetrating electron acceptor menadione, a membrane potential which was sensitive to rotenone was formed in both antimycin (+ cyanide)-treated mitochondria and particles. These data suggest that the membrane potential-forming mechanism of the respiratory chain is arranged in such a way that, on the outer surface of the mitochondrial membrane, there is no ferricyanide-reducing component before the site of antimycin inhibition.

According to Garland et al.[74], chlorquine, a non-penetrating agent which inhibits respiration of submitochondrial particles competitively with CoQ[74-76], is practically ineffective in intact mitochondria.

The above observations suggest that the respiratory chain segment between NADH and the antimycin-sensitive component is arranged along, rather than across, the membrane and is localised close to its inner surface. In this case, membrane potential generators of the first and second energy coupling sites should be organised similar to the transhydrogenase, i.e. as H^+ pumps. If this is the case, four out of five mitochondrial generators of membrane potential may be H^+ pumps and one (cytochrome oxidase) a transmembranous electron carrier.

2.3.5   Five electric generators in mitochondria: a general scheme

Figure 2.15 summarises the above concept of the energy coupling in mitochondria. It is proposed that the transfer of reducing equivalents between NADPH and the cytochrome c_1 region occurs close to the inner surface of the membrane. Then hydrogen atoms are transferred across the membrane to reach cytochrome c localised on the outer membrane surface. The transmembranous movement of hydrogen atoms may be catalysed by a cytochrome or a non-haeme protein if this electron carrier responds to reduction with such a change in the pK of one of its proton-accepting groups that addition of electron is accompanied by addition of H^+. These relationships were demonstrated, e.g. for cytochrome b[77]. If it is cytochrome b that is involved in the hydrogen atom transfer across the membrane, it should have a redox potential close to cytochrome c, i.e. about $+0.2$ V, to avoid energy dissipation at this stage of the respiratory chain. Cytochrome c_1 is characterised by this very potential value. However, it is not clear whether cytochrome c_1 is protonated when reduced.

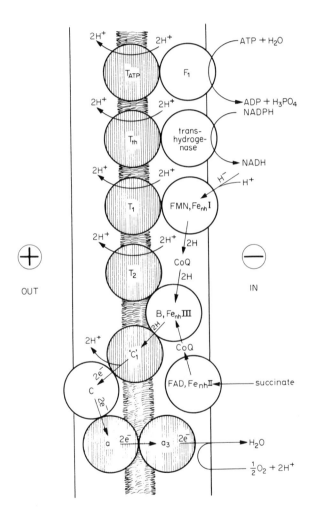

Figure 2.15 The general scheme of the membrane potential generators in mitochondrial membrane. H^+-translocating components are designed as T_{ATP} (ATPase), T_{th} (transhydrogenase), T_1 (1st coupling site) and T_2 (2nd coupling site); Fe_{nh}, non-haeme iron

In any case, cytochrome c is reduced on the outer surface of the mito-chondrial membrane. Subsequent transfer of an electron from cytochrome c, via cytochrome oxidase, to oxygen is directed across the membrane, so that oxygen reduction takes place on the inner surface of the membrane. Hence, the third energy coupling site is assumed to be an electrogenic oxidoreduction between cytochrome c and oxygen. The other three coupling sites, as well as ATPase, are postulated to be organised as H^+ pumps composed of a catalytic subunit responsible for an exergonic chemical reaction, and of an H^+-trans-locating system component in the uphill H^+ movement coupled with the energy-releasing reaction.

It should be noted that the above version of the respiratory chain is consis-tent with the lack of exchange of the transhydrogenase-carried hydrogen atoms with water, and with the initial stages of respiratory chain being located near the inner surface of the mitochondrial membrane. The scheme in question does not require alternations of the electron and hydrogen carriers in the chain since both H (or H^-) and e^- transfer, if coupled with the opera-tion of an H^+ pump, may form a membrane potential. The stoichiometry $H^+:e^-$ for the redox H^+ pump may differ from 1, depending on the amount of H^+ ions attached to the protonophorous subunit of the pump under the experimental conditions used.

Electric generators of the mitochondrial membrane are functioning revers-ibly (the only exception being, probably, oxygen reduction, the terminal step of the cytochrome oxidase reaction). Hence the membrane potential formed by one generator can be utilised by another to carry out chemical work: ATP synthesis or reverse electron transfer against a gradient of redox potential.

Apparently, ATP synthesis proves to be a more complicated function than transport of hydrogen atoms, H^- or e^-. This may be a reason for the specific knob-like organisation of the ATPase generator.

It is noteworthy that a freeze-fracture study of the inner mitochondrial membrane revealed the half-membrane facing the mitochondrial matrix to contain more particles (apparently, protein complexes) than the other half-membrane[78]. This fact is in agreement with the scheme given in Figure 2.5.

2.3.6 Essay on evolution of the membrane potential generators

2.3.6.1 *From photosynthesis to respiration*

It is most probable that various electric generators found in membranes of living organisms appeared at different steps of the biosphere evolution. One of the possible suggestions concerning development of such systems is based on the postulate that oxidoreduction between excited chlorophyll and an electron acceptor should have been the primary mechanism of charge separa-tion in biological membranes. To charge the membranes by this mechanism it would be sufficient to place chlorophyll and acceptor on the opposite sides of the hydrophobic barrier of the membrane. Transmembranous electron transfer from excited chlorophyll to the acceptor should stabilise the chlorophyll-absorbed energy in the form of a membrane potential†. The subsequent return of the reducing equivalent, transported by a hydrogen carrier, to the chlorophyll side of the membrane should have resulted in cyclisation of electron transfer and generation of an H^+ potential difference across the membrane. Such a process could have been the precursor of the cyclic electron transfer in photosynthetic bacteria.

The use of light energy, stabilised as $\Delta\Psi$, to support ion transport against concentration gradients could have been the first type of work performed by a membrane system. To do this work, rather simple ionophores, like valinomycin or nigericin, would have been required.

Appearance of a reversible H^+–ATPase complex incorporated in the same membrane should have been the next and very important step of evolution of $\Delta\mu_H$ generators. Reversal of the H^+–ATPase reaction at the expense of $\Delta\mu_H$ produced by the chlorophyll system should result in ATP synthesis from ADP and P_i.

Formation of one more $\Delta\mu_H$-generating mechanism in the same cyclic electron transfer chain should have been the next step in developing the set of energy transducers of photosynthetic bacteria. This new generator localised at the cytochrome b level could be organised like the chlorophyll system (i.e. $\Delta\mu_H$ formation by antiport of electrons and hydrogen atoms), or like the H^+–ATPase (i.e. transmembraneous movement of H^+ ions). The H^+–ATPase, which appeared as we postulated after the chlorophyll generator, seems to be a more perfect mechanism of $\Delta\mu_H$ production. In this case, electrochemical H^+ gradient is formed by a system (H^+–translocase) specialised only in H^+ ions transport. The catalytic reaction furnishing the energy for uphill H^+ movement is carried out by another system specialised in the ATP splitting (ATPase subunit). It is essential that these two parts, constituting the H^+–ATPase generator, could evolve independently. Not so the chlorophyll generator. In this case, membrane charging is the direct result of the catalytic reactions *per se*, so that the independent evolution of catalytic and translocating functions is impossible. It may also be important that organisation of a proton pump requires only one carrier, whereas that of the loop requires four redox carriers. Taking into account these considerations one may presume that the new (cytochrome b) generator was organised like the H^+–ATPase rather than the chlorophyll system. This may also be the case for the

† There are some indications that the oxidoreduction between chlorophyll and the primary electron acceptor in the chloroplast or chromatophore membrane occurs in the transverse direction. It was shown[79-85] that delayed fluorescence of chlorophyll emitted 1–10 ms after illumination requires maintenance of a membrane potential. Any agents discharging the membrane potential inhibited this emission. Under the same conditions, a discharge of the ΔpH was without effect. This means that the delayed (1–10 ms) fluorescence is a function of $\Delta\Psi$ and not of ΔpH or total $\Delta\mu_H$. These relationships could be expected if the delayed fluorescence is the result of the following events:

$$Chl^+ + Ae^- + \Delta\Psi \rightarrow Chl^* + A \qquad (2.6)$$

$$Chl^* + \rightarrow Chl + h\nu \qquad (2.7)$$

Reaction (2.6) represents electron transfer between chlorophyll and the primary electron acceptor. This process, being transversally directed in the membrane, can be supported by the transmembraneous electric field. Reaction (2.7) is light emission by excited chlorophyll. Protons are not involved in these processes, so the ΔpH should be ineffective. It may be recalled that reverse electron transfer from cytochrome a to c is also supported[31] by $\Delta\Psi$ and not ΔpH.

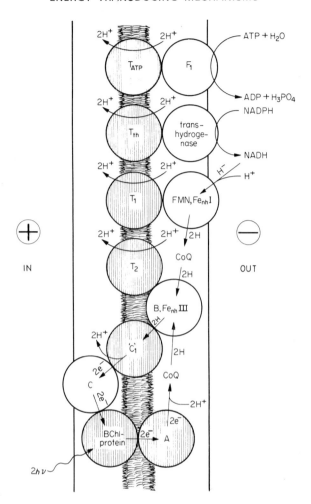

Figure 2.16 Membrane potential generators in the bacterial chromatophore membrane. BChl, bacterio-chlorophyll; A, primary electron acceptor; 'c_1', a c-type cytochrome which in chromatophores may perform the functions of cytochrome c_1

NADH dehydrogenase and transhydrogenase. Appearance of these two generators meant formation of a pathway for reduction of NAD^+ and $NADP^+$ by electron donors of positive redox potential, the process being carried out at the expense of light or ATP energy. Thus, formation of a system of membrane potential generation corresponding to that of photo-synthetic bacteria was completed (Figure 2.16).*

* It is highly probable that there is one more $\Delta\mu_H$ generator in the chromatophore membrane utilising inorganic pyrophosphate in an oligomycin-insensitive manner[18, 29]. The system of the H^+–pyrophosphatase could have been an evolutionary precursor of H^+–ATPase.

Development of non-cyclic photosynthesis including water photolysis could have been the result of the opening of the cyclic redox chain of photosynthetic bacteria. According to the hypothetical scheme shown in Figure 2.17, the cycle is opened in such a way that electrons are transferred from the primary electron acceptor of photosystem I (A_I) to ferredoxin (Fd) and then to $NADP^+$ via $NADP^+$–reductase flavoprotein. Cytochrome b is reduced by

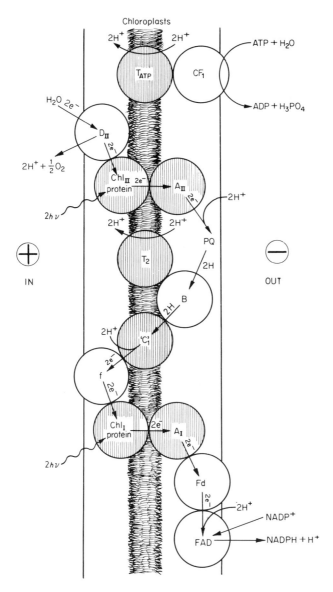

Figure 2.17 Membrane potential generators in the chloroplast membrane. Chl, chlorophyll; D_{II}, electron donor of photosystem II; PQ, plastoquinone; Fd, ferredoxin

plastoquinone (PQ) instead of CoQ. The primary electron acceptor of photosystem II (A_{II}) serves as a PQ reductant. The system of water photolysis functions as electron donor (D_{II}) for oxidised chlorophyll II. The chloroplast redox chain is postulated to include three $\Delta\mu_H$-generating steps, namely two electronic generators (steps $Chl_{II} \rightarrow A_{II}$ and $Chl_I \rightarrow A_I$) and one protonic localised at the cytochrome b level.

Accumulation of molecular oxygen gave rise to the appearance of a respiratory chain. The latter may also be considered as an open derivative of the cyclic chain of photosynthetic bacteria. The similarity of these two systems is really conspicuous. The only modification distinguishing the respiratory chain from the redox chain of photosynthetic bacteria is the replacement of the 'bacteriochlorophyll primary acceptor' by cytochrome oxidase (*cf.* Figures 2.15 and 2.16).

Primarily, the respiratory chain appeared, probably as an additional mechanism of energy conservation in some photosynthetic bacteria. Further specialisation of these organisms in the heterotrophic, light-independent way of life should have resulted in disappearance of the photosynthetic apparatus which was substituted by a respiratory chain as the only system of electron transfer phosphorylation.

2.3.6.2   *From bacteriorhodopsin to animal rhodopsin*

Among bacteria, one can find many combinations of photosynthetic, respiratory and glycolytic types of energetics. In this context we would like to discuss only one but striking example of this kind, namely halophilic bacteria.

We mentioned above (see p. 46) that these micro-organisms transform their energetics from aerobic heterotrophic to photosynthetic when the bacterial culture reaches the exponential growth phase. It was shown that light energy utilisation in *Halobacterium halobium* is catalysed by a retinal-containing rhodopsin-like protein (bacteriorhodopsin). Evidence that bacteriorhodopsin functions as an electrogenic light-dependent proton pump has been obtained. This system may be defined as 'evolutionary secondary photosynthesis.' It could have arisen as the result of adaptation of oxygen-consuming bacteria to the conditions in which the O_2 concentration, initially low due to high salt concentration, decreased further because of the respiration of densely populated cells.

The remarkable similarity between bacteriorhodopsin and animal rhodopsin in molecular weight, amino acid composition, and light-induced changes of properties of the retinal–protein complex can hardly be a coincidence. We should look for a common function and, maybe, for a common origin of these proteins in spite of the great distance between the halobacteria and higher animals in the evolutionary system.

The bacteriorhodopsin-mediated proton translocation should include two events: reversible protonation of bacteriorhodopsin and the channelling of H^+ ions across the membrane. One may assume that animal rhodopsin retained only one of these functions, namely the transmembrane channelling of ions which is no longer controlled by protonation–deprotonation reactions. If so, the light might induce a conformation change in rhodopsin resulting in formation of a thorough, ion-conducting channel in the membrane of photoreceptor discs of retina cell. Efflux of some ions (e.g. Ca^{2+})[86] via this channel may initiate a long chain of subsequent events resulting in visual excitation.

2.3.6.3   *From procaryotes to eucaryotes. Substitution of $\Delta\mu_H$ by $\Delta\mu_{Na}$*

Appearance of eucaryotic cells meant development of intracellular compartmentation of different metabolic functions. In particular, the function of energy production, associated with the cell membrane or its derivatives in procaryotes, was transferred to the membranes of specialised organelles, mitochondria and chloroplasts, an event which perhaps included a stage of symbiosis of eucaryotic and procaryotic cells.

The cell membrane of procaryotic organisms, in addition to its function in energy production, is responsible for transport of extracellular compounds from the medium into the cell. The latter function is supported by the same driving force as ATP synthesis, i.e. electrochemical potential of H^+ ions produced by a redox chain. The fact that it is the hydrogen ion whose gradient is formed by the redox chain is apparently a result of the chemical mechanism of the evolutionary primary reaction of the membraneous energy conservation organised as a light-induced e^-/H antiport.

There are several independent lines of evidence that in mitochondrial and chloroplast, as well as in bacterial membranes, $\Delta\mu_H$ (or one of its constituents, $\Delta\Psi$ or ΔpH) is the driving force not only for ATP synthesis but also for the uphill transport of different compounds (for a review, see Refs. 15, 22, 25, 29 and 87). To maintain $\Delta\mu_H$, the membranes in question must have very low H^+ permeability and high electric resistance. This is not an easy matter, especially if we take into account that the hydrophobic phospholipid bilayer of the membranes is interrupted by membraneous proteins of redox chain carriers, translocases, etc. It is not surprising, therefore, that minor changes in the native membrane structure result in the uncoupling of oxidation and phosphorylation and inhibition of transport processes.

There are numerous indications for a symport of Na^+ and extracellular compounds as a mechanism of transport processes in animal cell membranes (for a review, see Refs. 87 and 88). Such a symport was postulated to be catalysed by a carrier binding both Na^+ and the compound which accumulated in the cell. The ternary complex composed of a carrier, Na^+ and a transported compound moves into the cell down the Na^+ gradient, the free carrier returns down the carrier gradient, and Na^+ is pumped out by the Na^+,K^+–ATPase. This mechanism resembles the symport of H^+ and lactose into cells of *E. coli*[89, 90] or of H^+ and fatty acyl compounds (with carnitine as the carrier) into mitochondria[91]. It is remarkable that systems of this type can utilise both electric ($\Delta\Psi$) and concentration (ΔpH or ΔpNa) components of $\Delta\mu_H$ ($\Delta\mu_{Na}$). In other words, the high electric resistance of the membrane is not obligatory for the transport process if the membrane permeability for Na^+ is low.

Na^+ has an advantage over H^+ in that it is the commonest cation in the environment. Besides, the amount of substances which bind Na^+ is rather low, so that the buffer capacity of the extra- and intra-cellular medium for Na^+ is much lower than for H^+. As a result, it is much easier to obtain a high gradient of Na^+, than of H^+, pumping Na^+, instead of H^+, out of the cell.

It is impossible to reduce the concentration of Na^+ in the cell without substituting it by another cation. This cation proved to be K^+, the second widespread monovalent cation in the environment. Therefore the Na^+ pump could have been organised as a Na^+,K^+-antiport system with ATP as energy source. The mechanism of Na^+,K^+-ATPase may be essentially the same as that of the H^+-ATPase.

It seems attractive to consider the possibility that membraneous and non-membraneous energy-transducing mechanisms of the cell such as the actomyosin system are of common origin. This system, like the H^+-ATPase, can be resolved into two protein components, one required for catalysis (myosin) and another (actin) for the coupling of the catalytic reaction to the performance of work. According to the current concepts of muscle contraction, formation of a local electric field between one of the charged products of ATP hydrolysis (ADP or P_i) bound to actomyosin, and a fixed negative charge on this protein complex, is essential for the contractile function[92, 93].

Further comparative analysis of actomyosin and some membraneous ATPase seems to be desirable. It is noteworthy that ATP hydrolysis catalysed by both actomyosin and Ca^{2+}-ATPase of sarcoplasmic reticulum includes the same stages of the Mg^{2+}-dependent enzyme phosphorylation and Ca^{2+}-dependent decomposition of the phosphorylated enzyme.

One can hope that various energy-transducing catalysts of the cell will be described in terms of a universal concept. Such a concept may be based on the idea of the 'work-performing catalyst' put forward by Engelhardt[94] in 1944. Today, this idea may be developed as a hypothesis postulating a biological energy transducer consisting of two components, one responsible for catalysis of an exergonic reaction, and the other for performance of work. It is proposed that both components may have developed independently to form a perfect system whose functioning includes the following two main steps. (i) Energy-releasing transformation of a substrate into a product, accompanied by an energy-consuming transition of the enzyme–product complex into an energetically unfavourable conformation (e.g. the bringing together of similar electric charges). (ii) Conformational change resulting in conversion of the local electric field energy into useful work. This hypothesis has something in common with the general ideas developed recently by Green and associates[95]; however, the approach to the problem as well as the visualisation of energy coupling mechanisms in Green's and our concept are quite different.

2.4 CONCLUSIONS

2.4.1

Recent progress in oxidative phosphorylation research has resulted in the following fundamental findings consistent with Mitchell's chemiosmotic principle of energy coupling:

(1) *There are two types of enzymic systems (oxidative and dephosphorylative) which generate a membrane potential in mitochondria.* Proteins involved in the functioning of these two generators can be separated and reconstituted with phospholipids to form proteoliposomes competent in oxidation-, or, alternatively, ATPase-supported membrane potential formation. Transmembraneous electric potential difference, generated by both cytochrome oxidase and ATPase proteoliposomes, can be *directly measured by a voltmeter* under conditions favourable for fusion of proteoliposomes with planar phospholipid membrane.

Proteoliposomes containing both oxidative and ATPase enzyme systems generate a membrane potential at the expense of the reactions of both types:

oxidation → membrane potential ← ATP hydrolysis

and catalyse oxidative phosphorylation.

(2) *ATP-supported formation of a membrane potential is reversible*, so that the membrane potential can be utilised for ATP formation. This conclusion is supported by observations on ion transfer phosphorylation coupled with downhill ion fluxes of proper direction.

(3) *Any agents, discharging the membrane potential, uncouple oxidation and phosphorylation.* Such an effect can be obtained with (a) protonophorous uncouplers, (b) ionophorous antibiotics carrying out cyclic transport of K^+ across the membrane, and (c) systems generating a membrane potential of the polarity which is opposite to that required to ADP phosphorylation.

These facts seem to the author of this review to be sufficient to assume that the membrane potential is a factor in the coupling of oxidation and phosphorylation:

oxidation → membrane potential → ATP synthesis

2.4.2

The following observations are the most essential for the formulation of the mechanisms of membrane potential generation by the respiratory chain and the ATPase enzymes.

(1) *Proteins of the cytochrome oxidase complex and, very probably, the cytochrome oxidase-mediated electron transfer are organised across the mitochondrial membrane.* Thus, membrane charging in the third energy coupling site seems to be due to the transmembranous electron flow.

(2) *Electron transfer is not involved in the ATP-supported generation of membrane potential.* H^+, *instead of* e^-, *seems to be the charged species whose transport across the membrane is responsible for the membrane potential generation coupled with ATP-hydrolysis.* The mitochondrial enzyme complex, which catalyses the oligomycin-sensitive ATP dephosphorylation coupled with H^+ transport (H^+–ATPase), can be resolved into a catalytic component, F_1, hydrolysing ATP in the oligomycin-insensitive faction, and an oligomycin-sensitive H^+-translocating hydrophobic component of the mitochondrial membrane. The latter, being associated with F_1, is involved in the ATP ↔ membrane potential energy interconversion. Without F_1, the

H^+-translocating component acts as an oligomycin-sensitive protono-phorous uncoupler.

(3) *Hydrogen atoms transferred from one nicotinamide nucleotide to the other during mitochondrial transhydrogenase reaction are not exchanged with hydrogen ions of water.* The transhydrogenase reaction, even if it takes place across the membrane, cannot result directly in $\Delta\mu_H$, differing in this respect from the cytochrome oxidase system. Another difference is that both the hydrogen donor and the acceptor react with the transhydrogenase system on the same (matrix) side of mitochondrial membrane.

(4) *Ferricyanide does not interact with the respiratory chain before the cytochrome* c $(+ c_1)$ *step in intact mitochondria and does interact with inside-out submitochondrial particles.* It therefore seems probable that the respiratory carriers of the first and second energy coupling sites are localised on the matrix side of the membrane.

2.4.3

The available data may be explained in terms of the following concept of energy coupling.

(1) There is only one redox loop in the respiratory chain, corresponding to the third energy coupling site which forms $\Delta\mu_H$ as a result of hydrogen atom/electron antiport.

(2) The ATPase, the transhydrogenase, and the first and second energy coupling sites are organised as H^+ pumps carrying out transmembranous movement of H^+ against the H^+ electrochemical gradient.

(3) Each H^+ pump is composed of (a) a catalytic protein localised near the matrix surface of mitochondrial membrane; and (b) protein(s) organising a proton-conducting pathway across the middle (hydrophobic) part of the membrane.

(4) Generation of a membrane potential by both electron transfer system (cytochrome oxidase) and H^+ transfer systems (H^+ pumps) is postulated to include a stage of an energised conformation of the protein complex, e.g. appearance of neighbouring similar charges, so that the chemical energy of oxidative or dephosphorylative reaction is converted primarily into a local electric field energy. Electric repulsion of similar charges induces rotation (or some other conformational change) of a component transferring H^+ or e^- across the membrane. At this stage the local electric field is converted into the transmembranous electric field.

(5) From the evolutionary point of view, the primary mechanism of membrane charging may be visualised as photoinduced oxidoreduction between chlorophyll and an electron acceptor which are localised on the opposite sides of the membrane. This mechanism is retained, apparently, in the membrane potential generators of photosynthetic bacteria and chloro-plasts which are associated with chlorophyll, as well as in the cytochrome oxidase generator. Reversible $\Delta\mu_H$ generation by H^+ pumps specialised in the active H^+ transport seems to be a secondary, and more perfect, mechan-ism of membrane charging which is used in all generators other than those involving chlorophyll or cytochrome oxidase.

2.4.4

Among membrane potential generators, the simplest one is bacteriorhodopsin from halophilic bacteria. It is composed of retinal and polypeptide chain of mol. wt. 26 000 localised in special membrane areas containing no other proteins. Bacteriorhodopsin catalyses a light-dependent electrogenic H^+ transport which was demonstrated in intact cells, bacteriorhodopsin proteoliposomes and planar phospholipid membrane with bacteriorhodopsin inclusions. In the latter system, the light-induced electrogenesis has been directly measured with ordinary electrometer techniques. One can hope that further study on bacteriorhodopsin will allow the elucidation of the principles of action of biological electric generators.

Acknowledgements

The author is grateful to Professor P. Mitchell, Professor E. Racker, Professor S. Ye. Severin, Dr. Yu. V. Evtodienko, Dr. A. N. Glagolev, Dr. A. A. Jasaitis, Dr. A. A. Konstantinov, Dr. I. A. Kozlov, Dr. E. A. Liberman, and Dr. L. S. Yaguzhinsky for valuable advice, discussion and criticism; to Professor E. Racker and Professor W. Stoeckenius for information on unpublished results, and to Miss T. I. Kheifets for correcting the English version of the paper.

References

1. Mitchell, P. (1966). *Chemiosmotic Coupling in Oxidative and Photosynthetic Phosphorylation* (Bodmin: Glynn Research)
2. Kagawa, Y. and Racker, E. (1971). *J. Biol. Chem.*, **246**, 5477
3. Racker, E. and Kandrash, A. (1971). *J. Biol. Chem.*, **246**, 7069
4. Kagawa, Y. (1972). *Biochim. Biophys. Acta*, **265**, 297
5. Hinkle, P. C., Kim, J. J. and Racker, E. (1972). *J. Biol. Chem.*, **247**, 1338
6. Racker, E. (1972). *J. Membrane Biol.*, **10**, 221
7. Ragan, I. and Racker, E. (1973). *J. Biol. Chem.* **248**, 2563
8. Mitchell, P. and Moyle, J. (1969). *Europ. J. Biochem.*, **7**, 471
9. Liberman, E. A., Topali, V. P., Tsofina, L. M., Jasaitis, A. A. and Skulachev, V. P. (1969). *Nature*, **222**, 1076
10. Grinius, L. L., Jasaitis, A. A., Kadziauskas, J. P., Liberman, E. A., Skulachev, V. P., Topali, V. P., Tsofina, L. M. and Vladimirova, M. A. (1970). *Biochim. Biophys. Acta*, **216**, 1
11. Bakeeva, L. E., Grinius, L. L., Jasaitis, A. A., Kuliene, V. V., Levitsky, D. O., Liberman E. A., Severina, I. I. and Skulachev, V. P. (1970). *Biochim. Biophys. Acta*, **216**, 13
12. Isaev, P. I., Liberman, E. A., Samuilov, V. D., Skulachev, V. P. and Tsofina, L. M. (1970). *Biochim. Biophys. Acta*, **216**, 22
13. Liberman, E. A. and Skulachev, V. P. (1970). *Biochim. Biophys. Acta*, **216**, 30
14. Grinius, L. L., Il'ina, M. D., Mileykovskaya, E. I., Skulachev, V. P. and Tikhonova, G. V. (1972). *Biochim. Biophys. Acta*, **283**, 442
15. Skulachev, V. P. (1971). *Current Topics Bioenerg.*, **4**, 127
16. Liberman, E. A. and Topali, V. P. (1969). *Biofisika USSR*, **14**, 452
17. Hinkle, P. C. (1973). *Abstr. Conf. on Mechanism of Energy Transduction in Biol. Systems*, Abstr. No. 7 (New York: Academy of Science)

18. Barsky, E. L., Bonch-Osmolovskaya, E. A., Ostroumov, S. A., Samuilov, V. D. and Skulachev, V. P. (1975). *Biochim. Biophys. Acta*, (in the press)
19. Montal, M., Chance, B. and Lee, C.-P. (1969). *Biochim. Biophys. Res. Commun.*, **36**, 428
20. Montal, M., Chance, B. and Lee, C.-P. (1970). *J. Membrane Biol.*, **2**, 201
21. Papa, S., Guerieri, F., Lorusso, M. and Qiagliariello, E. (1970). *FEBS Lett.*, **10**, 295
22. Harold, F. M. (1972). *Bacteriol. Rev.*, **36**, 172
23. Drachev, L. A., Jasaitis, A. A., Kaulen, A. D., Kondrashin, A. A., La Van Chu, Semenov, A. Yu., Severina, I. I. and Skulachev, V. P. (1975). *J. Biol. Chem.*, (in the press)
24. Jasaitis, A. A., Nemeček, I. B., Severina, I. I., Skulachev, V. P. and Smirnova, S. M. (1972). *Biochim. Biophys. Acta*, **275**, 485
25. Skulachev, V. P. (1972). *FEBS Symp.*, **28**, 371
26. Skulachev, V. P. (1974). *Ann. N. Y. Acad. Sci.*, **227**, 188
26a. Drachev, L. A., Jasaitis, A. A., Mikelsaar, H., Nemeček, I. B., Semenov, A. Yu. Semenova, E. G., Severina, I. I. and Skulachev, V. P. (1975). *J. Biol. Chem.*, (in the press)
27. Racker, E. (1970). *Essays Biochem.*, **6**, 1
28. Schneider, D. L., Kagawa, Y. and Racker, E. (1972). *J. Biol. Chem.*, **247**, 4074
29. Skulachev, V. P. (1972). *Energy Transformation in Biomembranes*, (Moscow: Nauka Press)
30. Palmieri, F. and Klingenberg, M. (1967). *Europ. J. Biochem.*, **1**, 439
31. Hinkle, P. and Mitchell, P. (1970). *J. Bioenerg.*, **1**, 45
32. Wikström, M. K. F. (1973). *Abstr. Conf. on Mechanism of Energy Transduction in Biol. Systems*, Abstr. No. 6 (New York: Academy of Science)
33. Grinius, L. L. and Skulachev, V. P. (1971). *Biokhimiya USSR*, **36**, 430
34. Dontsov, A. E., Grinius, L. L., Jasaitis, A. A., Severina, I. I., and Skulachev, V. P. (1972). *J. Bioenerg.*, **3**, 277
35. Arion, W. J. and Racker, E. (1970). *J. Biol. Chem.*, **245**, 5186
36. Jasaitis, A. A., Severina, I. I., Skulachev, V. P. and Smirnova, S. M. (1972). *J. Bioenerg.*, **3**, 387
37. Groot, G. S. P., Kovač, L. and Schatz, G. (1971). *Proc. Nat. Acad. Sci. USA*, **68**, 308
38. Oesterhelt, D. and Stoeckenius, W. (1971). *Nature New Biol.*, **233**, 149
39. Blaurock, A. E. and Stoeckenius, W. (1971). *Nature New Biol.*, **233**, 152
40. Oesterhelt, D. (1972). *Z. Physiol. Chem.*, **353**, 1554
41. Oesterhelt, D. and Stoeckenius, W. (1973). *Proc. Nat. Acad. Sci. USA*, **70**, 2853
42. Racker, E. and Stoeckenius, W. (1974). *J. Biol. Chem.*, **249**, 662
43. Kayushin, L. P. and Skulachev, V. P. (1973). *FEBS Lett.*, **39**, 39
43a. Drachev, L. A., Kaulen, A. D., Ostroumov, S. A. and Skulachev, V. P. (1974). *FEBS Lett.*, **39**, 43
43b. Drachev, L. A., Frolov, V. N., Kaulen, A. D., Liberman, E. A., Ostroumov, S. A., Plakunova, V. G., Semenov, A. Yu. and Skulachev, V. P. (1975). *J. Biol. Chem.*, (in the press)
44. Garrahan, P. J. and Glynn, I. M. (1966). *Nature*, **211**, 1414
45. Jagendorf, A. T. and Uribe, E. (1966). *Proc. Nat. Acad. Sci. USA*, **55**, 170
46. Cockrell, R. S., Harris, E. J. and Pressman, B. C. (1967). *Nature*, **215**, 1487
47. Barlogie, B., Hasselbach, W. and Makinose, M. (1971). *FEBS Lett.*, **12**, 267
48. Makinose, M. (1971). *FEBS Lett.*, **12**, 269
49. Panet, R. and Selinger, Z. (1972). *Biochim. Biophys. Acta*, **255**, 34
50. Greville, G. D. (1969). *Current Topics Bioenerg.*, **3**, 1
51. Hinkle, P. C. and Horstman, L. L. (1971). *J. Biol. Chem.*, **246**, 6024
52. Slater, E. (1973). *9th Int. Biochem. Congr. Abstr.*, 8 (Stockholm)
53. Eley, D. D., Mayer, R. J. and Pething, R. (1973). *J. Bioenerg.*, **4**, 271
54. Wakabayashi, T., Senior, A. E., Hatase, O., Hayashi, H. and Green, D. E. (1972). *J. Bioenerg.*, **3**, 339
55. Junge, W. (1972). *FEBS Lett.*, **25**, 109
56. Schwab, A. I. and Weiss, H. (1972). *Abstr. 8th FEBS Meeting*, Abstr. No. 657 (Amsterdam)
57. Tzagoloff, A. (1973). *Biochim. Biophys. Acta*, **301**, 71

58. Poyton, R. O., Ross, E. and Schatz, G. (1973). *9th Int. Biochem. Congr. Abstr.*, 249 (Stockholm)
59. Mitchell, P. (1972). *J. Bioenerg.*, **3,** 5
60. Glagolev, A. N. and Skulachev, V. P. (1974). *Biokhimiya USSR,* **39,** 615
61. Ryrie, I. J. and Jagendorf, A. T. (1972). *J. Biol. Chem.*, **247,** 4453
62. Lee, C.-P. and Ernster, L. (1965). *Biochem. Biophys. Res. Commun.*, **18,** 523
63. Scholes, P., Mitchell, P. and Moyle, J. (1969). *Eur. J. Biochem.*, **8,** 450
64. House, D. R. and Packer, L. (1970). *J. Bioenerg.*, **1,** 273
65. Cattell, K. J., Lindop, C. R., Knight, J. C. and Beechey, R. B. (1971). *Biochem. J.*, **125,** 169
66. Stekhoven, F. S. (1972). *Biochem. Biophys. Res. Commun.*, **47,** 7
67. Severina, I. I. (1971). *Abstr. 7th FEBS Meeting*, 225 (Varna)
68. Skou, J. C. (1972). *FEBS Symp.*, **28,** 339
69. Liberman, E. A. and Tsofina, L. M. (1969). *Biofisika USSR,* **14,** 1017
70. Grinius, L. L., Guds, T. I. and Skulachev, V. P. (1971). *J. Bioenerg.*, **2,** 101
71. Lee, C. P., Simard-Duquesne, N., Ernster, L. and Hoberman, H. D. (1965). *Biochim. Biophys. Acta,* **105,** 397
72. Griffiths, D. E. and Roberton, A. M. (1966). *Biochim. Biophys. Acta,* **118,** 453
73. Orlando, J. A. (1968). *Arch. Biochem. Biophys.*, **124,** 413
74. Garland, P. B., Clegg, R. A., Downie, J. A., Gray, T. A., Lawford, H. G. and Skyrme, J. (1972). *FEBS Symp.*, **28,** 105
75. Skelton, F. S., Pardini, R. S., Heidker, J. C. and Folkers, K. (1968). *J. Amer. Chem. Soc.*, **90,** 5334
76. Ruzicka, F. J. and Crane, F. L. (1971). *Biochim. Biophys. Acta,* **226,** 221
77. Urban, P. F. and Klingenberg, M. (1969). *Europ. J. Biochem.*, **9,** 519
78. Melnick, R. L. and Packer, L. (1971). *Biochim. Biophys. Acta,* **253,** 503
79. Clayton, R. K. (1971). *Adv. Chem. Phys.*, **19,** 353
80. Fleischman, D. E. and Clayton, R. K. (1968). *Photochem. Photobiol.*, **8,** 287
81. Wright, C. A. and Crofts, A. R. (1971). *Europ. J. Biochem.*, **19,** 386
82. Barber, J. and Varley, W. J. (1971). *Nature New Biol.*, **234,** 188
83. Fleischman, D. E. (1971). *Photochem. Photobiol.*, **14,** 277
84. Crofts, A. R., Wright, C. A. and Fleischman, D. E. (1971). *FEBS Lett.*, **15,** 89
85. Sherman, L. A. (1972). *Biochim. Biophys. Acta,* **283,** 67
86. Bonting, S. L. (1969). *Current Topics Bioenerg.*, **3,** 351
87. Mitchell, P. (1970). *Symp. Sci. Gen. Microbiol.*, **20,** 121
88. Kotyk, A. (1973). *Biochim. Biophys. Acta. Rev. Biomembranes,* **300,** 183
89. West, I. C. (1970). *Biochim. Biophys. Res. Commun.*, **41,** 655
90. West, I. C. and Mitchell, P. (1972). *J. Bioenerg.*, **3,** 445
91. Levitsky, D. O. and Skulachev, V. P. (1972). *Biochim. Biophys. Acta,* **275,** 33
92. Davies, R. E. (1963). *Nature,* **199,** 1068
93. Bendall, J. R. (1969). *Muscles, Molecules and Movement* (London: Heinemann)
94. Engelhardt, W. A. (1944). *Isvest. Akad. Nauk SSR, Biol. Ser.*, **2,** 182
95. Green, D. E. and Sung Chul Ji. (1972). *J. Bioenerg.*, **3,** 159

3
Electrical Excitability in Lipid Bilayers and Cell Membranes

P. MUELLER

Eastern Pennsylvania Psychiatric Institute, Philadelphia

3.1 INTRODUCTION

The ionic conductance of many cell membranes is controlled by the membrane potential and if the effect is very conspicuous the cell is called electrically excitable. Voltage dependent conductances do not occur only in nerve and muscle but have also been found in primitive algae[1,2], many plant cells[3], protozoae[4], ovae[5,6] and in epithelial cells of higher animals[7].

If the transmembrane ionic gradients and the sequence of the conductance changes are appropriate, the membrane generates action potentials. In other cases there occur only regenerative transitions from one potential and conductance level to another[2], but in all instances the crucial event is a rapid, efficient and reversible change of the membrane permeability to ions in response to changes of the membrane potential.

We shall concern ourselves here mainly with the molecular aspects of this event, keeping in mind that some of the ideas might also apply to receptor and synaptic membranes in which other energy forms—light, heat, pressure or ligand binding—induce similar permeability changes.

It is now generally agreed that ions cross the membrane at localised, sparsely distributed regions thought of as pores or channels[8] and that these structures somehow are opened or closed, i.e. gated, by the membrane potential. In many instances the channels conduct certain ions preferentially at the partial or complete exclusion of others. In addition to the well known potassium and sodium channels of nerve and muscle, electrically gated channels for divalent cations[9] and anions[10] have been found in other cells. In some cases the ion specificity is poor[11] and the channels exclude only anions or cations but do not distinguish between the different ions in each class.

In contrast to the great differences in ion selectivity among different channels, their gating mechanisms seem to have a great deal in common. The steady state and kinetic control of the gating process by the membrane potential shows characteristic features which were first recorded and analysed in the squid axon[12-16] and have since been found to occur in essentially identical form in other cells and tissues. In all cases the basic aspects of the gating characteristics can be quantitatively described by the Hodgkin–Huxley (H & H) equations[16] provided only that the appropriate constants in these equations are properly adjusted.

The structural aspects of the cell membrane are now much better understood than at the time when Hodgkin and Huxley undertook their analysis. The basic membrane matrix is a lipid bilayer acting as an inert barrier for ions and non-electrolytes. The special transport and permeability pathways are formed, as far as is known, by proteins or glycoproteins inserted into the bilayer. The ratio of lipid to protein varies, but in most cases there is sufficient lipid to form a fluid matrix, allowing rapid lateral diffusion of both lipid and protein.

Lipid bilayer membranes separating two aqueous phases can be formed in planar and vesicular form by a variety of methods[17-23], and several compounds extracted from cells have been found[18, 24, 25] which, when incorporated into the bilayer, generate ion conducting pathways with voltage dependent gating characteristics indistinguishable from those found in excitable cells. In some cases the structure of these compounds is known and thus the minimal components and conditions for constructing prototype excitability mechanisms have been obtained. The gating kinetics displayed by these artificial systems match the H & H kinetics in every detail. The ion selectivities are less pronounced than in higher cells and generally restricted to cations or anions as a class. However, since these compounds come from primitive cells—bacteria and algae—whose inherent ion specificities are not known, and because the ion selectivity may be an issue separate from the gating mechanisms, the restricted selectivities should not distract from the possibility that the mechanism by which the membrane opens and closes for the passage of ions is essentially the same in artificial bilayers and in cells.

In the following, we should like to elaborate this point, present some data demonstrating the similarity of the gating kinetics in bilayers and cells, suggest a particular molecular model for the bilayer case and discuss its possible application to cellular excitability.

3.2 THE BILAYER

During the past decade much has been learned about the properties of lipid bilayers, partly owing to new methods for their formation allowing the study of single bilayers and partly owing to modern analytical techniques; although a detailed summary of the topic lies outside the scope of this review, we should like to point out a few facts of relevance to the later discussion.

First of all, it is important to realise that the bilayer in cells and artificial systems is a planar liquid, a fact that was already appreciated by the classical physiologists and that is now well established owing to the application of modern techniques, such as spin labelling[26, 27], n.m.r.[28], fluorescence spectroscopy[29-31] and calorimetry[32]. Typical lateral diffusion coefficients are in the order of 10^{-8} cm^2 s^{-1} and are influenced by head group composition, hydrocarbon chain saturation, ionic strength of the aqueous phase and temperature.

The second point is that the thickness of the hydrocarbon region as determined by x-ray diffraction in a variety of cells and artificial membranes is only close to 30 Å, i.e. much less than the combined length of the two opposing hydrocarbon chains[19, 33-36]. This is so because the polar head groups of the phospholipids generally occupy a larger area than the extended hydrocarbon chains, so that in order to maintain their normal density the chains must bend and the membrane thickness decrease. Finally, the apparent dielectric thickness of the hydrocarbon region as determined from the electrical capacity is even smaller than the structural thickness, i.e. of the order[19] of 20 Å. This discrepancy can be readily understood by assuming that some water molecules are present in the region of the ester groups and penetrate even somewhat beyond into the hydrocarbons. The presence of water would make this region more conductive and this may also at least partially account for the observed frequency dependence of the capacity. Recent e.s.r. measurements have confirmed this suggestion[37].

In most of the experiments on artificial systems discussed below, the bilayers were formed from lipids dissolved in hydrocarbon solvents. These membranes are somewhat thicker but also more fluid due to the retention of some solvent. Controls with solvent free membranes[38] show only minor quantitative differences in terms of gating rates, single channel lifetimes and membrane sensitivity to a particular translocator, but no new phenomena have been observed.

3.3   EXCITABILITY INDUCING TRANSLOCATORS

So far, four different compounds have been found to induce electrical excitability: EIM is a protein obtained from enterobacter cloacae[23]; alamethicin a cyclic peptide from Trichoderma Viride[25, 39]; monazomycin is a small molecule (mol. wt. 1400) possibly related to the polyene antibiotics and is produced by a streptomyces strain[40]; DJ400B is a true cyclic polyene[41].

Of the four, only the structure of alamethicin[42] and DJ400B[41] are known in detail and the molecular mechanism was derived on the basis of their structure, but the similarity of the gating characteristics and some of the known molecular properties of EIM and monazomycin give reason to believe that the proposed principles apply to all four compounds.

3.4   CHARACTERISTICS OF THE VOLTAGE DEPENDENT
CONDUCTANCE CHANGES

The main steady state and kinetic features of the gating process as they appear from voltage clamp studies in the excitable bilayers and cells are presented in Figures 3.1–3.4. These features are accounted for by the H & H

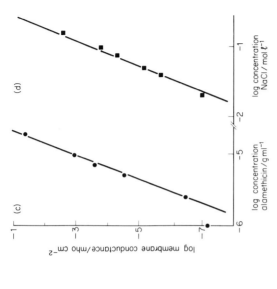

Figure 3.1 (a) The relation between the membrane potential and the potassium conductance in nerve. (From data in **Ref. 14.**)

(b) The same relation in a lipid bilayer (mixed brain lipids–cholesterol) in the presence of 2×10^{-6} M monazomycin on one side and 0.1 M KCl on both sides. Note the different positions of the curves in (a) and (b) with respect to the voltage axis. The g–V curve for alamethicin has about the same position as that of monazomycin, whereas with EIM in sphingomyelin membranes the curve is positioned as in nerve. In this and all following bilayer records the sign of the indicated potentials is that of the compartment containing the translocator

(c) **Relation between the membrane conductance and the concentration of alamethicin in the aqueous phase at constant ion concentration (0.1 M NaCl) and constant voltage (60 mV)**

(d) **Relation between the membrane conductance and the concentration of NaCl in the aqueous phase at constant alamethicin concentration (10^{-5} g ml^{-1}) and constant voltage (60 mV). Oxidised cholesterol membrane**

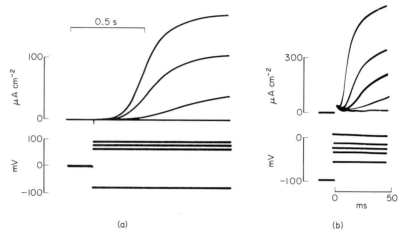

(a)   (b)

Figure 3.2 (a) Membrane currents in response to applied potential steps of different amplitude and sign for a bilayer (oleyl phosphate–cholesterol) in the presence of 10^{-6} M monazomycin on one side and 0.01 M NaCl on both sides; temperature 25 °C. Several oscillograph sweeps are superimposed. Note the long delay of the current rise and the absence of measurable current for negative potential steps

(b) Membrane currents in response to potential steps for the potassium system in muscle. The sodium currents were blocked by tetrodotoxin. (From Adrian et al.[50], by courtesy of Cambridge University Press.)

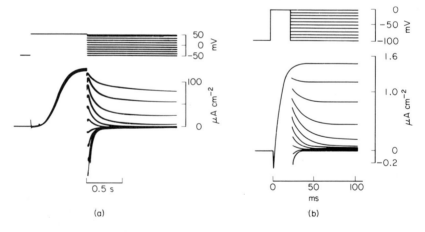

(a)   (b)

Figure 3.3 (a) Membrane currents in response to applied potential steps for a bilayer in the presence of 10^{-6} M monazomycin (one side) and 0.01 M KCl (both sides). Eleven records are superimposed. The voltage steps were applied at 20 s intervals from a holding potential of -50 mV. Note that the time constants of the exponential conductance decays show a maximum near 35 mV (third trace from top in the current records). The brief current maximum at the beginning of the negative going voltage steps are a phenomenon separate from the exponential current decay and are discussed in Section 3.8.2.3

(b) Similar data as in (a), but for the potassium system in muscle. The negative spike near zero time in the current record is due to the transient increase of the sodium conductance and obscures the delayed rise of the potassium currents visible in Figure 3.2b. (From Adrian et al.[50], by courtesy of Cambridge University Press.)

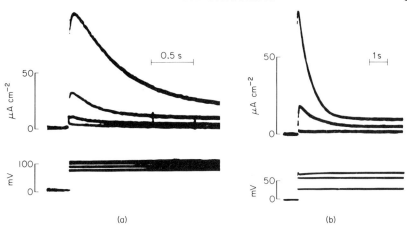

(a)   (b)

Figure 3.4 (a) Membrane currents in response to applied potential steps from a bilayer (glycerol diolein–lecithin) in the presence of 10^{-5} M monazomycin and 0.001 M KCl; temperature 38 °C. Because of the low ionic strength, the fluid lipid (high glycerol diolein) and the high temperature, the initial delay of the current rise is very short and not visible on this timescale. Note that the current increases and then declines (inactivation) even though the potential is held constant. Similar records but on a different timescale have been obtained with alamethicin and EIM

(b) Inactivation in the potassium system of muscle. (From Adrian *et al.*[50], by courtesy of Cambridge University Press.)

analysis, which implies that the time course of the conductance change depends only on the membrane potential and time, but not on the past history and present state of the gating mechanism. There appear, under certain conditions, deviations from this simple state of affairs but their discussion is left until later.

3.4.1   The steady state conductance–voltage relation

Several aspects of the steady state conductance–voltage relation (Figures 3.1a, 3.1b) should be noted. At low conductance levels the curve is exponential and the conductance changes e-fold for a potential change of 5–8 mV. At higher potentials the conductance reaches a plateau. The curves can be shifted parallel to the voltage axis by agents affecting the surface potential such as multivalent ions[43, 44], local anaesthetics[45], the ionic concentrations[46, 47] and, in the case of the bilayer, also by the nature of the lipid head groups[24] and the identity of the channel forming compound[18, 48].

In the artificial systems the sign of the rectification, i.e. the sign of the voltage resulting in a conductance increase, is determined by the side to which the translocator is added; in all cases seen so far, a positive potential applied to that side increases the conductance. Increasing concentrations of the translocators result in a shift of the curve parallel to the conductance axis, and the conductance measured at a fixed potential increases with the nth power of the concentration, with values of n depending on the nature of the translocator and the membrane lipid composition. Typically, values of n range from 5 to 10 for alamethicin and monazomycin[18, 25], less for EIM (Figure 3.1c). In the case of alamethicin the conductance at a given potential is also a high power function of the salt concentration (Figure 3.1d), and the slope of the conductance–voltage curve is proportional to the valency of the cation in the aqueous phase[49].

3.4.2   Kinetic characteristics

The important kinetic details are shown in Figures 3.2–3.4.

3.4.2.1   The delayed conductance increase and voltage dependent time constants

The transition of the conductance from a low to a high level in response to an applied potential step proceeds along an S-shaped curve with an initial delay that is more or less pronounced in different cells and that depends in the bilayer on the translocator, the membrane lipids and the ionic composition of the aqueous phase (Figure 3.2). The delay becomes shorter as the size of the voltage step is increased and disappears when the initial resting conductance is raised either by the potential level from which the step is applied (holding potential) or, in the bilayer case, by increasing the concentration of the translocators.

When the sign of the applied voltage is such that it decreases the conductance, the time course of the decrease is exponential and the time constants go through a maximum as a function of the potential (Figure 3.3)[16].

3.4.2.2   Inactivation

In the sodium channels of nerve the conductance increase after a voltage step of appropriate sign goes through a maximum in time, a phenomenon called inactivation. Although inactivation is most conspicuous in the sodium system, it is not restricted to these channels, appearing to a varying degree and depending on conditions in all channel types[50-54]. It can also be demonstrated in the excitable bilayers with EIM, alamethicin and monazomycin[18, 55] (Figure 3.4). In the bilayers the ratio of the peak to the steady state conductance increases with increasing concentration of the translocators, the temperature and, in the case of alamethicin, with the Ca concentration in the aqueous medium[38]. The degree to which the membrane inactivates can be measured by two different methods: one is the evaluation of the peak to steady state ratio of the conductance change for different voltage steps; the other is the determination of the ratio of the peak conductance obtained by a voltage step from a holding potential at which the conductance is very low to the peak conductance obtained by a voltage step from different holding potentials at which the conductance is higher. The curves of inactivation vs. membrane potential obtained by the two methods are slightly different but generally of the same shape[56].

3.5 THE HODGKIN–HUXLEY EQUATIONS

The steady state and kinetic gating characteristics, summarised briefly in the preceding paragraphs, form the essential background for the H & H analysis and since the bilayers show the same characteristics, it is not surprising that their gating process can also be described by the H & H equations[55].

Hodgkin and Huxley designed their equations primarily as a mathematical description of the conductance changes without placing undue emphasis on a molecular interpretation. Nevertheless, they discussed several possibilities, among them the suggestion that the voltage controls the position of charged particles normal to the plane of the membrane and that several of them have to occupy a particular position in order to form a bridge or channel for the flow of ions. Neglecting for the moment the inactivation process, their formulation reduces to the simple scheme (3.1) and (3.2):

$$P_0 \underset{k_{10}}{\overset{k_{01}}{\rightleftharpoons}} P_1 \tag{3.1}$$

$$g \approx (P_1)^n \tag{3.2}$$

Here P_0 represents the fraction of charged particles at one side of the membrane, P_1 those on the other side and the voltage dependent rate constants k_{01} and k_{10} are of the general form:

$$k_{01,10} = A \exp\left[-(E_{01,10} \mp 0.5\, zV\, 23.05)/RT\right] \tag{3.3}$$

in which A is a constant and E_{01} and E_{10} represent the energy barriers the particles have to cross in order to go from position 0 to position 1 and back, V the membrane potential and z the valency of the particles. The sign of their charge determines only the vectorial aspects, i.e. which side of the membrane the particles occupy in the closed state and what sign the voltage must have in order to open the channels. E is expressed in kcal mol^{-1} and V in eV. Because several P_1 particles are needed to open one channel, the conductance g is assumed to be proportional to the nth power of P_1; n has usually a value of 3 or 4. In the steady state, i.e. at a fixed holding potential, V_H:

$$P_1 = \frac{k_{01}^H}{k_{01}^H + k_{10}^H} \tag{3.4}$$

In response to a sudden displacement of the potential to a new value, V_C, P_1 changes according to:

$$\frac{dP_1}{dt} = k_{01}^C(1 - P_1) - k_{10}^C P_1 \tag{3.5}$$

Hence

$$P_1 = \frac{k_{01}^C}{k_{01}^C + k_{10}^C} - \left(\frac{k_{01}^C}{k_{01}^C + k_{10}^C} - \frac{k_{01}^H}{k_{01}^H - k_{10}^H}\right) \exp\left[-t/(k_{10}^C + k_{10}^C)\right] \tag{3.6}$$

where the superscripts H and C denote the rate constants at the holding, V_H, and clamp potential, V_C.

In the original H & H formulation, P_1 is labelled n or m for the K or Na system and the rate constants are called a and β. Their form differs somewhat from the straight Boltzmann relation of equation (3.3), but this was done only to allow a better fit to the data. If the E values in equation (3.3) are chosen properly, the particles are mainly at one side of the membrane at the normal resting potential of nerve and are moved by a depolarising potential towards the other side where several of them are assumed to interact or aggregate and allow the flow of ions.

This system accounts not only for the steady state conductance–voltage curves, but the equations (3.5), (3.6) and (3.2) describing the changes of P_1 and P_0, and g as a function of voltage and time, also fit the kinetic data, i.e. the delayed conductance increase, the exponential decay and the variation of the time constants as a function of the membrane potential.

In order to account for the inactivation, Hodgkin and Huxley had to assume a separate process which, obeying a similar but first order mathematical description, would either prevent the channels from opening or close them after they had opened.

Leaving this particular problem for later, we should like to emphasise at this point the extraordinary success these simple equations have enjoyed in describing membrane excitability phenomena. In practically every instance where sufficient data have been accumulated and where it has been possible to separate the conductance changes of one system of channels from other systems being also operative in the same membrane, the H & H equations have been able to accommodate the data with only minor adjustment of the constants.

Moreover, the basic concepts involved have recently received additional support from the discovery of the so-called gating currents[57-59], which are an implicit consequence of the scheme and had been predicted by H & H. These currents are generated by the movement of the charged particles, P_0 and P_1 in (3.1) in response to the applied field. In theory they should be smaller, faster and occur earlier than the conductance change and follow Boltzmann kinetics, and although some quantitative deviations from the minimal H & H relations seem to occur, all in all the predictions have been borne out in a most remarkable way.

It is thus only reasonable to ask if indeed there is a common molecular mechanism of electrical excitability and if a molecular interpretation of the H & H equations together with insights gained from the bilayer systems can give clues to its nature.

3.6 MOLECULAR ASPECTS OF THE GATING PROCESS

3.6.1 Two alternatives: configurational change or aggregation

The delayed conductance changes, the exponential dependence of the steady state conductance on the voltage and, in the bilayer case, on the translocator concentration all speak strongly for a cooperative phenomenon. In principle there are two different mechanisms compatible with this notion.

The first involves a configurational change within a fixed preformed channel structure. The second, a voltage dependent insertion of channel precursors into the membrane and their subsequent aggregation into a functional channel.

In the first case the preformed channel must undergo a configurational change in response to an applied field and the observed high order effects would result from internal cooperative interactions, perhaps a helix coil transition or an allosteric interaction between channel subunits, similar to the interactions in regulatory enzymes. Such a mechanism is quite conceivable but the specific assumptions that would have to be made in order to arrive at a molecular model are, at the present state of knowledge, quite arbitrary. This is especially true for the sodium and potassium system of nerve and muscle where information about the identity and structure of the channel forming molecules is still lacking.

In the bilayers the situation is different. The structures of alamethicin and DJ400B are known and some information has been obtained for monazomycin and EIM. Furthermore, the experimental control of the lipid composition, and of the concentration, orientation and chemical nature of the translocators has provided additional data not yet obtainable in the cellular systems. All of this information taken together points, at least in the bilayer case, strongly to the second alternative and a mechanism has been proposed[25] that is compatible with the original H & H suggestion and their equations. It is assumed that in the resting state, i.e. when the membrane conductance is low, the translocators lie at the membrane surface, held there by interactions with the lipid polar groups and that an applied field drives them into the membrane hydrocarbon region where they aggregate by lateral diffusion to form channels.

3.6.2 The aggregation scheme

The aggregation process can be described by the following scheme[60]:

$$P_0 \underset{k_{10}}{\overset{k_{01}}{\rightleftharpoons}} P_1$$

$$P_1 + P_n \underset{k_{(n+1)(n)}}{\overset{k_{(n)(n+1)}}{\rightleftharpoons}} P_{n+1} \tag{3.7}$$

$$P_1 + P_{m-1} \underset{k_{(m)(m-1)}}{\overset{k_{(m-1)(m)}}{\rightleftharpoons}} P_m$$

where P_0 represents the concentration of monomers at the surface, P_1 that of inserted monomers, P_2 to P_n that of the dimers, trimers, etc., P_m being the largest n-mer. The total concentration of translocators, C, is:

$$C = P_0 + \sum_{n=1}^{m} n P_n \tag{3.8}$$

The rate constants controlling the insertion of monomers k_{01} and k_{10} are assumed to be voltage dependent and of the same form as in equation (3.3).

The other rate constants are of the form:

$$k_{(n)(n+1)} = A \exp\left[-E_{(n)(n+1)}/RT\right] \qquad (3.9)$$

$$k_{(n+1)(n)} = A \exp\left[-E_{(n+1)(n)}/RT\right] \qquad (3.10)$$

In first approximation they are not voltage-dependent and are determined by the interaction energies and diffusion rates of the various fractions.

The concentration P_n of the different n-mers as a function of time and voltage can be derived by numerical methods from the differential rate equations of the reaction sequence (3.7) and the membrane conductance can be calculated by summing the conductances attributed to the individual oligomers. When the aggregation rates are much higher than the insertion rate, and only one oligomer configuration (e.g. the tetramer) is energetically preferred (but less preferred than the monomer), the formulation reduces to the H & H equations. Under these conditions the time constants of the observed clamping currents are only dependent on the field and not on the state of the system. This insertion–aggregation process may seem somewhat awkward and in this simple form may not be applicable to cells, but in the bilayer system it is supported by a great deal of almost compelling evidence.

3.6.3   Data supporting aggregation

Leaving the molecular details as far as they are known until later, we should like to list some of the supporting data and arguments, focusing mainly on the artificial bilayer systems.

(1) The conductance is generally a high-order function of the translocator concentration. This effect was already mentioned and is illustrated in Figure 3.1c. It indicates that more than one molecule is involved in the formation of a channel and has also been observed with channel formers that are not voltage dependent, such as nystatin[61], prymnesin[63] or gramicidin[64, 65]. There is, however, no direct theoretical correlation between the number of molecules forming a channel and the exponent relating the conductance to the concentration of the translocator in the aqueous phase. For one, the concentration of the translocator at the membrane surface is related to its concentration in the aqueous phase by a non-linear adsorption isotherm, and, secondly, the relation between the surface concentration and the concentrations of the various aggregates depends on their individual free energies. As a result, there exists no direct relation between the concentration of monomeric translocators in the aqueous phase and the concentration of oligomeric channels in the bilayer[61], and experimentally it is observed that the conductance–concentration function is a variable depending on lipids, ions and the identity of the translocator. Nevertheless, power functions with exponents greater than one and as high as 10 have been observed[25, 38, 61].

(2) The time constants of the conductance changes decrease with the

translocator concentration (Figure 3.5). This effect is quite pronounced and would be a direct consequence of the aggregation mechanism. As with the concentration dependence of the steady state conductance, so also in this case are the quantitative aspects dependent on the free energies of interaction between the translocator molecules. This observation is a rather strong argument for an aggregation mechanism. The only alternative but not very likely explanation would be an increase of the lipid fluidity caused by the translocators, which in turn would allow a more rapid opening of a pre-formed channel.

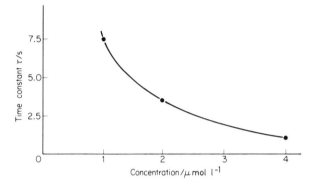

Figure 3.5 The dependence of the time constants of the conductance changes on the translocator concentration. The curve was obtained from an oleyl phosphate–cholesterol bilayer in the presence of the indicated concentrations of monazomycin (one side) and 0.1 M NaCl (both sides). The time constants were measured from the conductance increase in response to an applied potential step from − 10 to + 100 mV (From Banmann and Mueller[72])

(3) The time constants of the gating process depend on the lipid fluidity (see Figure 3.6). With alamethicin, the time constants in a very liquid lipid such as glycerol diolein[60, 62] are as small as 200 μs, i.e. comparable with those of the Na system in nerve, whereas in a more viscous membrane containing a high concentration of oxidised cholesterol, time constants of several seconds are the rule. Similar effects have been seen with monazo-mycin[55]. The phenomenon may not be entirely due to the lipid viscosity but could be partially caused by the low surface polarity of the glycerol diolein membranes, providing less interaction with the translocator and allowing a faster insertion by the applied field. In this context it is interesting that excitable cell membranes have an unusually high content of polyunsaturated hydrocarbon chains which could presumably result in patches of membrane with a low viscosity favouring short gating time constants.

(4) In the closed state, i.e. at high negative potentials, the translocators are assumed to lie at the membrane surface and therefore the membrane conductance in that state should be as low as that of the unmodified membrane, i.e. $10^8 \, \Omega \, cm^2$. This is in fact so, provided that the translocator concentration is not too high. In cells such high resistances at negative

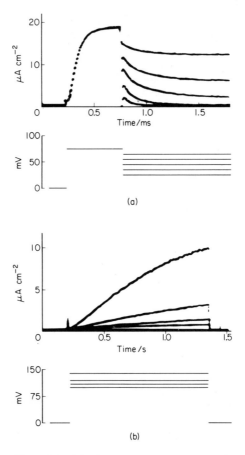

Figure 3.6 Membrane currents (top) in response to potential steps (bottom) for two different membranes in the presence of ala-methicin. The records in (a) were obtained from a membrane containing a high percentage of glycerol diolein (glycerol diolein/diolein phosphate/octane in volume ratios of 0.2:0.05:1). The membrane-forming solution in (b) contained 40 mg oxidised cholesterol and 50 mg dodecyl phosphate in 1 ml of octane. Several oscillograph sweeps were superimposed at 20 s intervals. Although the time scales differ by a factor of 1000 in the two records, the general sigmoid shape of the currents is preserved. (a) also demonstrates the exponential decay of the currents in response to a potential decrease and the variation of the time constants with the potential. The time constants in (a) are comparable with those in nerve. (From Baumann and Mueller[60], by courtesy of Alan R. Liss)

potentials are not observed because there are other leakage channels present in parallel with the excitable channels.

(5) For a particular translocator, the position of the conductance–voltage curve with respect to the voltage axis is determined by the chemical nature of the lipid polar groups and the surface potential of the bilayer–water interface. Conditions favouring the interaction of the translocator with the membrane surface tend to keep the gates closed, and, therefore, shift the conductance voltage ($g-V$) curve towards more positive potentials. For example, when lecithin is replaced by sphingomyelin, which provides additional hydrogen bonding groups, membrane potentials of almost 500 mV are required to induce a measurable conductance increase with monazomycin[38], and similar but less pronounced effects are seen with EIM[17, 24]. Conversely, the addition of protamine or other multivalent cations lowers the potential needed to increase the conductance. Shifts of the $g-V$ curve by agents affecting the surface potential are also well known in nerve[43], and the effects support the contention that in the closed state the parts of the translocators responsible for the gating rest at the membrane surface.

The other variable controlling the position of the $g-V$ curve is the chemical composition of the translocator as far as it determines self-interactions and interactions with the lipid hydrocarbons. If these interactions are strong, the equilibrium would be shifted towards the aggregates, i.e. channels, and at zero membrane potential the conductance would be high even at low translocator concentration. All known channel formers, whether voltage dependent or not, are practically insoluble in pure hydrocarbon solvents and show strong aggregation tendencies in aqueous solutions. EIM, which has probably a large monomeric molecular weight, forms aggregates of up to several million dalton[66] and like the K or Na system of nerve is conductive at zero membrane potential, whereas the smaller translocators need either a positive potential or must be present at high concentrations before the conductance increases measurably above the background of the unmodified membrane.

It is therefore apparent that the position of the steady state $g-V$ curve depends on a balance of interactions: those that favour the aggregation of the translocators within the bilayer phase and those that hold the translocator at the membrane surface.

(6) Several kinetic effects are not accounted for by the H & H equations, but nevertheless provide additional support for an insertion–aggregation mechanism. As already mentioned, the H & H assumption that the conductance is directly proportional to the nth power of the inserted translocators renders all time constants dependent only on the voltage but not on the state of the system, i.e. the time constants are neither dependent on the number of channels that happen to be open at the start of a voltage step, nor on the rate and sign of the conductance change when the voltage step is applied from a non-steady-state. This is so because the statement $g \approx (P_1)^n$, if interpreted as an aggregation mechanism, assumes implicitly that the rate of aggregation is infinitely fast, compared with the insertion rate, so that only the latter is rate limiting. In this sense the H & H equations are an incomplete description of an aggregation process and can be expected to predict the data accurately only when the aggregation rates are much faster

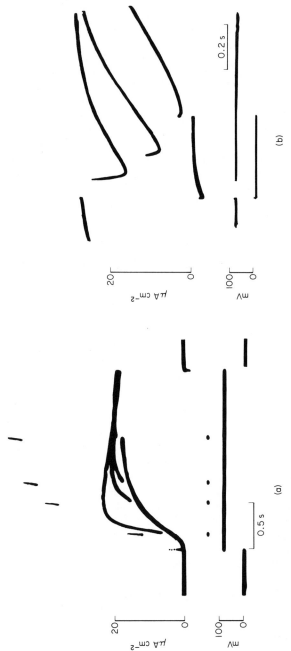

Figure 3.7 (a) Membrane currents (top) in response to applied potential steps (bottom) from an oxidised cholesterol–oleyl phosphate bilayer in the presence of 10^{-6} M monazomycin and 0.01 M KCl in the aqueous phase. Four records were superimposed. The brief voltage pulses to 150 mV were superimposed on a longer step of 90 mV at different times during the rise of the membrane conductance. The brief current increments during these pulses are seen as short spikes above the continuous current traces. At the end of the short voltage pulse the current continues to rise, reaching values well above the steady state level corresponding to the 90 mV potential of the longer step

(b) Conditions as in (a), except that three voltage pulses of -10 mV and different durations were applied at 20 s intervals from a clamp potential of 80 mV, i.e. a level at which the membrane conductance is high. The conductance falls during the -10 mV voltage step, as seen from the negative current trace at the bottom and the instantaneous values of the current at the end of each step, when the potential is returned to the 80 mV level. Note that after repolarisation the current continues to decline for a while before increasing again to the level corresponding to the 80 mV potential

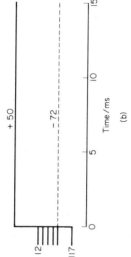

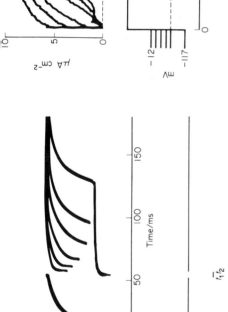

(a)

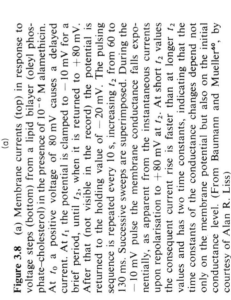

(b)

Figure 3.8 (a) Membrane currents (top) in response to voltage steps (bottom) from a lipid bilayer (oleyl phosphate-cholesterol) in the presence of 10^{-6} M alamethicin. At t_0 a positive voltage of 80 mV causes a delayed current. At t_1 the potential is clamped to -10 mV for a brief period, until t_2, when it is returned to $+80$ mV. After that (not visible in the record) the potential is returned to the holding value of -20 mV. The pulsing sequence is repeated every 10 s, increasing t_2 from 60 to 130 ms. Successive sweeps are superimposed. During the -10 mV pulse the membrane conductance falls exponentially, as apparent from the instantaneous currents upon repolarisation to $+80$ mV at t_2. At short t_2 values the consequent current rise is faster than at longer t_2 values and has two time constants, indicating that the time constants of the conductance changes depend not only on the membrane potential but also on the initial conductance level. (From Baumann and Mueller[60], by courtesy of Alan R. Liss)

(b) A similar experiment from the potassium system of frog nerve. In this case a voltage step to 50 mV was applied from different holding potentials between -117 to -12 mV. At the more negative holding potentials the membrane conductance is low and the current rises with the usual delay and a long time constant (the first two current traces from the bottom). As the initial conductance is increased at the less negative holding potentials, the currents rise without delay and their time constants becomes progressively shorter. In contrast to similar data from squid nerve[108], the current traces cannot be superimposed by a shift along the time axis and, as in (a), the time constants depend on both the voltage and the conductance. (From Hille[111], by courtesy of the Rockefeller Institute Press)

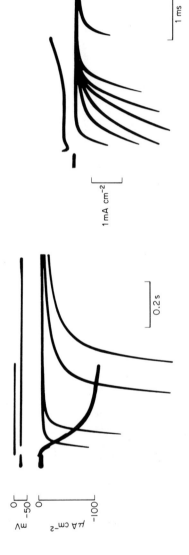

(c)

(d)

Figure 3.8 *continued*

(c) Membrane currents (bottom) in response to voltage steps (top) from a lecithin–cholesterol bilayer in the presence of 2×10^{-6} M monazomycin on one side. The aqueous phase contained 0.001 M KCl on the side of monazomycin and 0.05 M KCl on the opposite side. Four voltage steps of increasing duration were applied at 20 s intervals from a holding potential of -30 mV. The traces were superimposed. As the conductance increases during the voltage steps, the current flows towards the compartment containing the monazomycin. The large current 'tails' at the end of the voltage steps represent the decay of the conductance to its low level at the holding potential. Note that the time constants of the decay increase with the value of the conductance at the end of the voltage step

(d) Demonstrates the same phenomenon for the sodium system in nerve. In this case the voltage steps had a value close to that of the sodium equilibrium potential so that little current flowed during the step, but the time course of the conductance can be seen from the peaks of the current tails at the end of the voltage steps. Note that the conductance inactivates and that the time constants of the current tails depend, as in (a), on the conductance level at the time of repolarisation to the holding potential. The voltage steps are now shown but had the same pattern as in (a), except that seven steps of different duration were applied and the resultant traces superimposed. (From Frankenhaeuser and Hodgkin[112], by courtesy of Cambridge University Press)

than the insertion rates. In nerve this condition seems to be fulfilled to some extent, but it is also true that extensive tests of this point have not been made. The best known experiments of this type[108] have shown that in the K system of the squid axon the time course of the conductance increase caused by voltage steps from different holding potentials but going to the same clamp potential, can be superimposed after a shift along the time axis, as predicted from the H & H equations. Superposition—as this phenomenon is called—can also be demonstrated in the bilayers, provided that the lipid fluidity is high. However, the same experiments also showed that at high negative holding potentials the initial delay of the current rise became so long that the exponential factor n in equation (3.2) had to be raised from 4 in the H & H equation to 24 in order to fit the curves, i.e. the apparent cooperativity of the gating process became unusually large. Such long delays are normally seen with monazomycin (Figure 3.2), and as discussed below can be accounted for by an aggregation process involving only 5 or 6 monomers per channel, provided the aggregation rate constants are properly adjusted.

The superposition of the conductance time course in response to voltage steps starting from different steady state holding potentials is not a very sensitive test for revealing rate limiting steps of the gating reaction that are independent of the voltage. Better results are obtained if the gating time course is tested from different non-steady-state starting conditions. For example, if the conductance is changing in response to a voltage step and a second step in the opposite direction is applied before the conductance reaches the steady state, the conductance change may proceed for a short time in the original direction before inverting its sign. The effect can be demonstrated for increasing (Figure 3.7a) as well as decreasing (Figure 3.7b) conductances. In either case, the system behaves as if it possesses inertia and the gating process proceeds for a short time independent of the membrane potential. Another phenomenon of this type is shown in Figure 3.8.

These experiments demonstrate the dependence of the time constants on the conductance level and the past history of the gating system. The effects can be observed with all four translocators and are apparently also present in nerve (Figure 3.8b, d). However, it must be emphasised that they are variable, and occur in the bilayers only in the more viscous lipids, where the aggregation rates could be expected to become rate limiting. They disappear at high temperature and are not apparent in fluid lipids such as glycerol diolein where the kinetics are in strict accordance with the H & H equations.

3.6.4 A specific molecular model

Although none of the arguments listed above is compelling, considered together they provide suggestive evidence for an aggregation mechanism.

Further support comes from a consideration of the translocator structures and special effects such as single channel kinetics, and it has been possible to propose a detailed molecular model[60] of the gating process for alamethicin and DJ400B which may also apply to monazomycin and in modified form to EIM.

3.6.4.1 *Structural characteristics of a gating translocator*

The critical features of the model are most easily appreciated by considering the structure of DJ400B (Figure 3.9) and the scheme of Figure 3.10. They can be listed as follows:

1. The translocator molecule either is, or contains, an elongated structure long enough to span the hydrocarbon region of the bilayer.

2. At one end it must contain a hydrophilic group which either by its size or polar nature tends to anchor that part of the molecule at the lipid –water interface.

3. The other end should contain one or more charges and if the hydrophilic group at the opposite end is also charged, the two charged groups must have an opposite sign, giving the molecule a large dipole moment.

4. The elongated part of the molecule must contain at least several polar groups such as hydroxyl or carbonyl oxygens, facing to one side; the other side should be largely hydrophobic.

5. The molecules must aggregate in hydrocarbon solvents.

If these requirements are met, the gating is assumed to proceed in the following way.

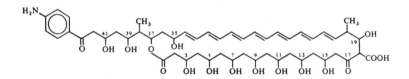

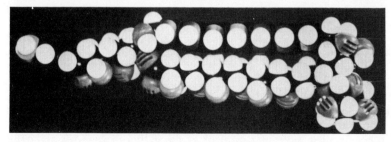

Figure 3.9 The structure and a molecular model of the cyclic polyene DJ400B. The formula (top) is that of the aglycone[41]. The position of the sugar in the model is hypothetical but in keeping with the usual position of this residue in other polyenes

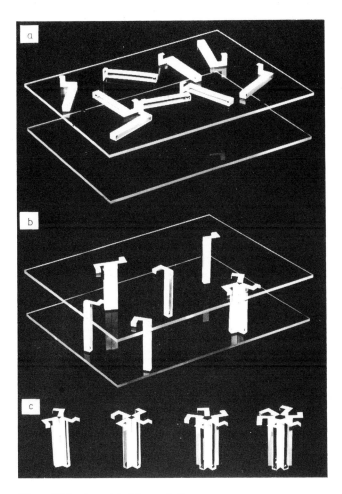

Figure 3.10 A model of the excitation process

(a) At rest, models of the translocator molecules lie flat on the membrane surface, represented by the upper plane

(b) An applied field acting on the charge, represented by the ball at one end of the model, has pulled the charged end of the translocator molecules into the membrane towards the *trans* surface

(c) Lateral diffusion within the membrane leads to aggregation of the monomers into oligomers. Trimers, tetramers, pentamers and hexamers form a central opening acting as a channel for the flow of ions

3.6.4.2   The gating process

At rest the molecules are assumed to lie flat on the membrane surface. An applied field of appropriate sign pulls the charged end through the membrane to the other side. The molecules now span the membrane with their long axis normal to the membrane plane. Lateral diffusion leads to aggregation into dimers, trimers, etc. Monomers and dimers do not conduct ions because they do not form an open channel, but higher aggregates do. In the higher aggregates the polar carbonyls or hydroxyl groups face towards the centre, forming the hydrophilic lining of the channel, whereas the non-polar sides of the translocators face towards the outside, interacting with the hydrocarbon chains. Upon removal of the field, the system returns to the original state.

In this model a considerable portion of the aggregation energy may be derived from the interaction of water with the hydrophilic groups of the channel interior, and this energy may also contribute to the configuration of the translocator. For this reason, a particular configuration in an aqueous environment may no longer be preserved after the translocators are inserted into the hydrocarbon region and have formed a channel. This consideration may apply especially to peptides such as alamethicin or EIM.

3.6.4.3   Specific examples

The conditions listed under 1–5 above can be fulfilled by a variety of different molecular structures. The cyclic polyenes are ideally suited to form this type of channel. Most of them do not have the required bipolar structure and the conductances they induce are therefore not voltage dependent. But even these compounds, e.g. nystatin and amphotericin, seem to form aggregates and channels of the 'barrel stave' type in which the molecules are oriented in such a way that their long axis spans the membrane[61]. The polyenes, including DJ400B, require cholesterol for their activity which may become an integral part of the channel structure, being inserted as a wedge between the polyene chains, thereby forcing them to line up in a circle. Monazomycin seems to be related to the polyenes having a similar molecular weight, i.r. and fluorescence spectrum[38,40] and chemical composition, and would, therefore, also fulfil the above conditions.

Similar 'barrel stave' channels may also be formed by peptide chains either singly or as hairpin loops[109] and the latter possibility may have been realised by alamethicin. In this case the cyclic peptide can be folded into the elongated ellipsoid shown in Figure 3.11 with one side essentially in β configuration, its peptide carbonyls providing the hydrophilic channel lining, the other side containing several hydrophobic α helix turns. The polar group attaching the molecule to the surface (condition 2, above) is formed by two glutamyl residues at one end and the gating charge by an inorganic cation held in fourfold coordination at the opposite end, in a cavity formed by the peptide carbonyls of the hairpin bend.

Since the gating charge is in this case not an inherent part of the molecule, the conductance should be and is the same high power function of the metal

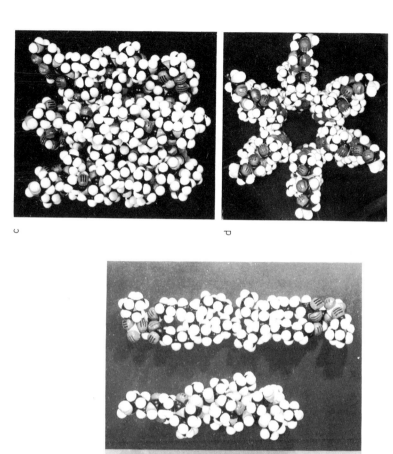

Figure 3.11 (a) The structure of alamethicin[42]. (b) Left and centre: A proposed molecular configuration of alamethicin, viewed from two sides. The metal ion is visible at the lower end of the molecule at the left (arrow). Right: Two lecithin molecules in bilayer configuration

(c + d) A molecular model of hexameric channel formed by the aggregation of six alamethicin molecules: (c), side view; (d), top view (From Baumann and Mueller[60], by courtesy of Alan R. Liss)

ion concentration as that of the alamethicin (Figure 3.1d). If the ion is divalent, the applied voltage would be twice as effective and the conductance–voltage relation is therefore twice as steep[49].

Very little is known about the structural aspects of EIM, but it should be emphasised that in sphingomyelin membranes its kinetics can show all the details discussed in Sections 3.4.2 and 3.6.3.

It tends to form large aggregates in aqueous solutions and its monomeric molecular weight, currently estimated as 25 000 dalton[66], is much larger than that of the other three translocators. This does, however, not need to exclude an aggregation mechanism, and it may be that only parts of the molecule—possibly short peptide chains—are inserted into the membrane by the field.

3.6.4.4 Additional support for the model

(a) *Single channel experiments*—Additional evidence for the aggregation mechanism comes from single channel experiments with alamethicin. Because the resistance of the unmodified bilayer is so high, it is possible to observe the conductance changes caused by the opening and closing of single channels. Such events have been recorded with EIM[17, 67, 68], alamethicin[69, 70] and gramicidin[71]. With alamethicin the conductance steps occur in bursts, the conductance switching rapidly between five to six levels. The spacing of the levels increases from the lower to the higher levels and the intermediate levels are most frequently occupied.

This bunching of the conductance steps, the increasing step size and the preferred occupation of the intermediate levels are a direct consequence of the aggregation mechanism. In fact, the values of the conductance levels have been quantitatively calculated from the molecular dimensions of the oligomeric alamethicin channels, assuming only that the specific conductivity with the individual channel is close to that of the aqueous phase[60] (Figure 3.12).

(b) *Polarisation of the gating mechanism*—The model is also supported by the apparent polarisation of the gating mechanism in a direction normal to the membrane plane. If the translocators are added to one membrane side, potentials of only one sign increase the conductance and when the applied field is removed, the system returns to its original state. If the translocators are added to both sides, two separate conducting systems result, each being gated independently by fields of opposite sign. The effect is a direct consequence of the model because as long as one end of the translocator stays firmly attached to one membrane surface the molecule will return to that side after the field is removed.

3.7 QUANTITATIVE CALCULATIONS

3.7.1 Basic kinetics

The reaction scheme accounts suprisingly well not only for the basic H & H steady state and kinetic data of Section 3.4—which should be expected—but

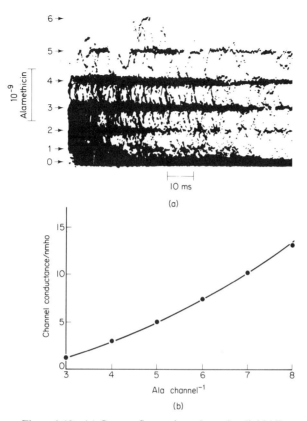

(a)

(b)

Figure 3.12 (a) Current fluctuations through a lipid bilayer membrane (glycerol monoleate) in the presence of 10^{-8} M alamethicin and 2 M KCl. The membrane potential was clamped at 210 mV. Many sweeps were superimposed on the screen of a storage oscilloscope, each sweep being triggered by the leading edge of a current transition. The baseline corresponds to the conductance of an unmodified bilayer. Six different current levels can be distinguished (arrows) and their relative probability of occurrence can be estimated from the intensity of the baseline traces. The third and fourth level are the most probable. The spacing between the levels increases with the current. (From Gordon and Haydon[69], by courtesy of Elsevier.)

(b) Correlation between the conductance levels in (a) and the conductance of hypothetical channels formed by the aggregation of 3–8 alamethicin molecules. The points represent the conductance levels and are calculated from the current levels in (a). The lowest level (1) is assumed to correspond to the conductance of a trimer. The curve represents the conductance of single channels built from 3–8 alamethicin molecules as shown in Figure 3.11 and was calculated from the molecular dimensions of the channel and the conductance of the aqueous phase. (From Baumann and Mueller[60], by courtesy of Alan R. Liss)

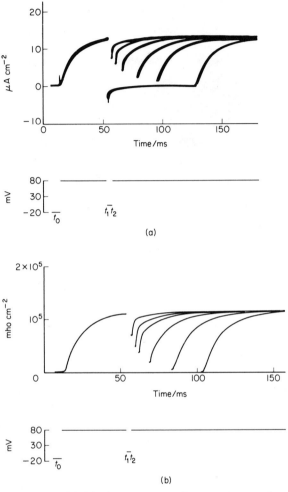

Figure 3.13 (a) Membrane currents in response to voltage steps from a bilayer in the presence of 10^{-6} M alamethicin. For details see Figure 3.8a

(b) Calculated conductance changes for the same pulse sequence as in (a). The curves were calculated from numerical solutions of the differential rate equations for the reaction scheme (3.7), assuming that the largest n-mer is a hexamer and that the conductances of the individual oligomers are as in Figure 3.12b. For clarity the conductance decay during the $+10$ mV pulse is not drawn. For further details see Ref. 60. In the model, the state dependence of the time constants results mainly from the retention of non-conducting channel precursors, i.e. monomers and dimers, in the hydrocarbon region during the $+10$ mV pulse. (From Baumann and Mueller[60], by courtesy of Alan R. Liss)

also for the deviations mentioned above in Section 3.6.3. Numerical solutions of the differential rate equations allow quantitative calculations of the channel concentrations as a function of voltage and time[60]. These calculations have shown[60, 72] that phenomena such as seen in Figures 3.7 and 3.8, as well as the variable delays and state dependent time constants, follow directly provided the individual rate constants are appropriately adjusted. An example is shown in Figure 3.13.

It could be argued that because of the large number of rate constants their individual values are rather arbitrary and almost any data could be matched. This is, however, not so. True, the fit of secondary details requires somewhat more carefully selected constants, but the basic phenomena appear with any reasonably realistic set. In fact, the simplest case, where the rate constants of the aggregation process are all equal, displays already some non-H & H effects, and variations of the rate constants only control quantitatively how much the kinetics deviate from the ideal H & H pattern, but do not introduce phenomena that are at variance with the experimental observations.

3.7.2 Inactivation

Inactivation, as already mentioned, occurs to a variable degree in the excitable bilayers and in cells. Because it is so generally present, it may be an inherent property of the gating mechanism and its general presence might give clues about the nature of the gating mechanism. In order to account for it, H & H assumed a separate, voltage dependent process governed by an independent set of equations. In recent years, theoretical arguments and experimental data have been presented tending to show that the inactivation is directly and sequentially coupled to the activation, i.e. to the channel opening[73, 74]. It is therefore of interest that the simple aggregation mechanism as shown in equation (3.7) can account for inactivation without any further ancillary assumptions, the only provision being that the rate constants are adjusted such that non-conductive members of the aggregation chain, e.g. the dimers, or trimers have a lower free energy and/or that their formation be slower—owing to lower rate constants—than that of the higher oligomers.

With these energetic restrictions, the concentration of higher oligomers, i.e. of conducting channels, builds up initially very fast from the inserted monomers, but then decays back into non-conductive dimers or trimers, and in the inactivated state the majority of the translocators are in the form of these lower aggregates.

Under such conditions it is possible to generate the entire set of kinetic phenomena related to inactivation, including both the basic H & H kinetics and deviations such as the delayed onset of inactivation[56] and the shifts of the inactivation curves along the voltage axis as a function of different test potentials. Some calculations are shown in Figures 3.14, 3.15 and 3.16. In this context, the inactivation could be looked upon as an aggregation or polymerisation overshoot, because the concentration of higher oligomers goes through a maximum as a function of time. The phenomenon is well known for linear polymerisations[75] and has been described and analysed for

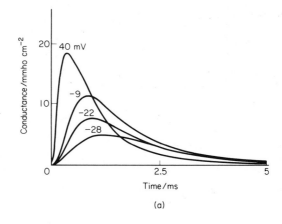

(a)

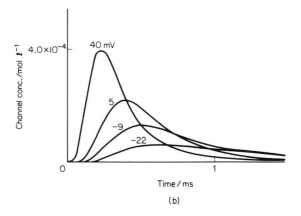

(b)

Figure 3.14 Inactivation in nerve (a) and in the aggregation model (b). The curves in (a) were plotted from data in Ref. 16 and represent the time course of the sodium conductance in the squid axon in response to voltage steps from a holding potential of $-60\,mV$ to the different potentials indicated above each curve. The curves in (b) represent the concentration of hexameric channels as a function of time in response to voltage steps from a holding potential of $-58\,mV$ to the indicated values. They were obtained from numerical solutions of the differential equations describing the aggregation scheme (3.7), assuming that only hexamers form a conductive channel and that they are the largest possible oligomers. The activation energies, E, of the rate constants [equations (3.9) and (3.10)] had the following values (in kcal mol^{-1}): $E_1 = 0.9$; $E_{10} = 2.1$; $E_{12} = 0$; $E_{21} = 3.2$; $E_{23}-E_{65} = 0$. All pre-exponential factors (A) were $= 2 \times 10^4$ and $C = 1$ mol l^{-1}. With this set of rate constants, the dimers are preferred and in the inactivated state most of the inserted monomers have formed non-conducting dimers. (From Baumann and Mueller[72])

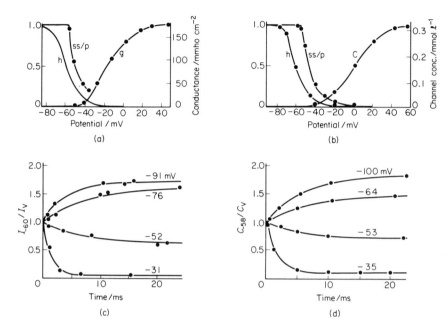

Figure 3.15 (a) The relation of the peak sodium conductance (curve g) and the inactivation (curves h and ss/p) to the membrane potential in the squid axon. Curve h represents the peak sodium currents in response to voltage steps from different holding potentials to a fixed test potential divided by the peak current in response to a voltage step from -80 mV to the test potential. The curve ss/p represents the ratio of the steady state to the peak conductance for responses to voltage steps from a holding potential of -80 mV to different test potentials (From data in Ref. 16).

(b) The relation between the peak hexameric channel concentration and the same voltage parameters as in (a), as calculated from the aggregation model. The rate constants were the same as in Figure 3.14b. (From Baumann and Mueller[72])

(c) The time course of the inactivation at four different membrane potentials in the sodium system of the squid axon. The ordinate represents the ratio of the peak sodium currents in response to voltage steps from different holding potentials (I_V) to the peak current from a holding potential of -60 mV (I_{-60}). The test potential was -16 mV, and the abscissa represents the time after application of the holding potential. (Replotted from Hodgkin and Huxley[15], by courtesy of Cambridge University Press)

(d) The relation between the peak channel concentration and similar voltage and time parameters as in (b), calculated from the aggregation model. The rate constants were as in (b). Note that there is a small delay at the onset and removal of inactivation, and that the time constant does not decrease at the most negative potential, as it does in (c). (From Baumann and Mueller[72])

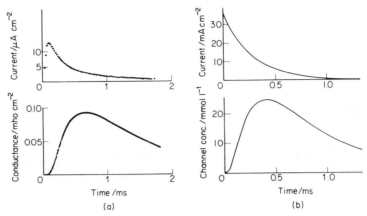

Figure 3.16 (a) The relation between the gating current (upper record) and sodium conductance (lower record) in the squid axon Na system. (From Armstrong and Bezanilla[57], by courtesy of Macmillan Journals)

(b) The same relation as calculated from the model. The calculations were done as in Figure 3.14b except that only the hexamer was assumed to be conductive. The lower curve represents the number of hexameric channels as a function of time. The upper curve shows the current due to the entry of monomers into the membrane. It was obtained by multiplying $-dP_0/dt$ by the Faraday constant. The potential step was from -80 to $+60$ mV. The following constants were used: $C = 1$, $A = 3 \times 10^6$, $z = 2$, $E_{01} = 5.4$, $E_{10} = 6.6$, $E_{12} = 3$, $E_{21} = 7$, $E_{23} = 1.5$, $E_{32} = 1.5$, $E_{34} = 1$, $E_{43} = 2.5$, $E_{45} = 0.5$, $E_{54} = 0.5$, $E_{56} = 0$, $E_{65} = 1.7$ (From Baumann and Mueller[60], by courtesy of Alan R. Liss)

the polymerisation of TMV coat protein[76] and a polynucleotide phosphorylase[77]. Of course it is not the only possible cause of inactivation. Blocking of the open channel by independent polypeptide or lipid molecules[97] or, in the case of monazomycin and alamethicin, the transmembrane diffusion of the entire molecule and concomitant depletion of monomers available for channel formation, are alternative mechanisms that must be considered[113].

3.8 FURTHER DETAILS OF THE INSERTION–AGGREGATION MECHANISM

Some consequences, details and modifications of the insertion and aggregation process need further consideration.

3.8.1 Some details of the insertion process

3.8.1.1 Gating currents

The recently reported gating currents in the Na system of the squid axon and their relation to the membrane potential, to the conductance changes and to the inactivation process agree quite well with the postulated insertion mechanism (see Figure 3.16). Such currents have not yet been observed in the bilayer systems, mainly because specific toxins which in nerve are used to block the much larger ionic currents are not available. However, the movement of charge associated with the tilting of the translocator into the membrane should give rise to such currents and the currents should have a particular relation to the holding potential and the amplitude as well as the direction of the applied potential steps. In the limited, strictly H & H case, the currents are described by equation (3.5). They should be exponential with a single time constant and their time integral [equation (3.6)] as a function of the clamp potential amplitude should be S shaped and decrease with increasingly positive holding potentials. This straightforward H & H scheme implies that the molecular entities carrying the gating charge are not involved in any secondary interactions and that their movement is only controlled by the membrane potential. In the more complex aggregation scheme (3.7), the gating charges become part of the channel structure and their movement under the influence of the applied field is modified by the interactions of the translocators with themselves and the lipids. Thus certain details of the gating current behaviour could give information about the aggregation process.

For example, it can be expected that the time constants of the gating currents connected with the closing of the gates become longer as more channels open up, because the translocators have aggregated and cannot be pulled back to the surface in the aggregated state. Therefore, the rates of disaggregation into monomers could become the rate limiting step for the removal of the monomers to the surface, i.e. the step $P_1 \rightarrow P_0$ in scheme (3.7).

The gating currents in nerve show this effect. Their time constants are exponentials, depend on the holding and clamp potentials in the required manner, but the time constants of the closing currents are longer if the gates are being closed at a time when the sodium conductance is at its peak rather than at earlier or later times[78, 79]. Obviously this phenomenon does not prove an aggregation process, but it does indicate that there is a time or state dependent process involving the gating structure that proceeds after the gating charges have been redistributed by the field.

If the movement of the gating charges would follow strict Boltzmann statistics, the charge of the individual component moving in the field could be determined from measurements of the integral of the gating currents as a function of the clamp potential. This interpretation however, assumes, first, that the gating charge moves through the entire potential drop across the membrane, and second, that the movement proceeds from one energy well at one side of the membrane over a hill to another well at the other side.

The first assumption may well be justified. In the bilayer case the translocator molecules are long enough to span the lipid hydrocarbon region and their postulated motion (Figure 3.10) would allow the gating charge to cross the hydrocarbon region and thereby utilise the entire energy of the applied field. This point is important if the system is to operate at peak efficiency.

The second assumption, however, needs a more careful consideration. The position of the gating charge at any point in the membrane profile is not only determined by the field, but also by the interaction energies of the translocator with the membrane lipids. If, for example, the gating charge were attached to the end of a highly non-polar peptide chain, its preferred position at zero potential might well be inside the hydrocarbon region and it might take work, i.e. an applied field, to pull the charge to the surface. In order to cross from a position at one membrane surface, corresponding to a closed channel, to a position at the other side the charge would pass through an energy valley instead of over a hill (Figure 3.17). Thus the Boltzmann relation (3.5) no longer would be valid and in rough approximation can be replaced by an expression accounting for the distribution of the gating charges between

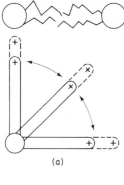

(a)

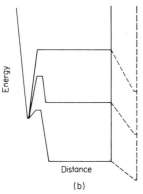

Distance

(b)

Figure 3.17 (a) A schematic representation of the insertion of a translocator molecule into the bilayer. The average orientation of the translocator axis with respect to the membrane plane depends on the membrane potential profile and the interactions of the translocator body with the aqueous phase and the lipid hydrocarbons

(b) Three hypothetical energy profiles as a function of the translocator orientation. The abscissa represents the position of the gating charge as it crosses the membrane. The profiles are drawn for zero membrane potential, and would be modified by the presence of a potential. The diagram illustrates two effects. (1) If the translocator is long enough for the gating charge to reach the opposite polar region [dotted lines in (a)], the interactions of the charge with the polar lipid groups and the water in that region give rise to an energy well (dashed lines), which is absent if the translocator is shorter [solid lines in (a) and (b)]. (2) If the parts of the translocator body extending into the hydrocarbon phase are sufficiently non-polar, the orientation normal to the membrane plane may have a lower free energy than the parallel orientation, as indicated by the bottom curve. In this case, the majority of the translocators are available for channel formation at zero membrane potential and it takes a negative membrane potential to pull them to the surface. If, on the other hand, the interaction of the translocator body with the hydrocarbons is small, it takes work to pull the gating charge away from the surface into the membrane, as indicated by the top curve. The middle curve represents an intermediate situation

three different membrane regions, the *cis* surface, the hydrocarbon region and the *trans* surface. Without going into the mathematical detail[106] it can be stated that, with increasing interactions of the translocator with the membrane hydrocarbons, the curves of gating current integrals as a function of the membrane potential become less steep so that the apparent gating charge is smaller. At the same time, the conductance–voltage curves are shifted towards more positive potentials, i.e. the channels tend to be open at zero potential. This situation may apply to EIM and perhaps to nerve, where an estimate of the gating charge on the basis of a Boltzmann relation gave a value of 1.3 electronic charges per gating particle[78] whereas in reality the gating charge may be divalent, perhaps a chelated Ca ion.

3.8.1.2 Effects of the translocator length

Another factor deserving consideration is the length of the inserted part of the translocator molecule compared with the thickness of the hydrocarbon region. If it were considerably shorter, the open channel would not reach to the other side and would be blocked by the lipids in the other half of the bilayer. In this case two half-channels inserted from opposite sides might meet and form an open channel, as may be the case with nystatin and gramicidin. On the other hand, if the molecules are too long, the gating charge can reach completely to the other side and interact with the polar groups and the aqueous phase. There would then be a large energy well, holding the molecule in a position spanning the membrane, favouring aggregation and keeping the channels open. This energy well would also tend to lower the rate constants for the removal of the molecules to the surface so that the closing of the channels by an applied field becomes very slow. DJ400B may fall into this class. It is $\sim$37 Å long and its amino group could reach the opposite membrane–water interface. Its closing rates are slow and a significant part of the channels, once opened by an applied field, do not close upon repolarisation. The ideal length seems to lie between the two extremes. The hydrophobic portion of alamethicin in the proposed configuration (Figure 3.11) is about 5 Å shorter than the lipid hydrocarbon region, but long enough for the open channel not to be blocked by the lipids. Because the gating charge stays in the hydrocarbon region, there is no energy well to hold the molecule in the inserted position and the gating is rapidly reversible.

3.8.1.3 The movement of aggregates in the field

The reaction scheme (3.7) assumes explicitly that only the monomers of the translocators are moved by the field, an assumption for which there is no evidence one way or another. In principle, dimers and higher oligomers may also be pulled out to the surface, but it can be argued that the larger the aggregate, the slower and less likely this effect would be. Quantitative calculations have not been made on this point. The gating currents and the conductance changes would be somewhat affected, but the overall gating characteristics would not change drastically.

There are indeed some experimental observations suggesting that the field can change the position of an entire channel within the membrane.

In membranes high in tocopherol or oxidised cholesterol, EIM forms open channels at zero membrane potential and these channels can be partially closed by potentials of either sign[17]. The conductance decreases, however, only 5- to 8-fold and the clamping currents give no indication of high-order effects. This partial and symmetrical gating, which has also been described for the K system in nerve[80, 107], stands in contrast to the complete gating in sphingomyelin membranes, where the conductance varies over 4–5 orders of magnitude and shows the typical delayed and high-order kinetics[55]. The most direct explanation for the partial gating would be that the voltage pulls the channels as a whole in either direction part way out of the membrane so that the lipid head groups can close the channel entrance to some extent.

3.8.2   Some details of the aggregation process

3.8.2.1   Different aggregation schemes

In the reaction scheme (3.7) it was assumed that the aggregation process is linear, i.e. proceeds sequentially to dimers, trimers, etc., by the addition of monomers, and that the oligomers themselves do not aggregate as units. Thus two dimers are not allowed to form a tetramer. This assumption, however justified it may be in each particular case, is only of importance in connection with the inactivation process because only if the aggregation of oligomers is negligible does the reaction lead to an overshoot of the concentration of higher aggregates. The other kinetic phenomena are unchanged.

The conductance steps seen with alamethicin indicate that there are channels of different diameters depending on the number of monomers forming the channel and that either the trimer or the tetramer gives the lowest measurable conductance. This need not always be the case. The relative stability of the different n-mers may be different in a different lipid environment or for a different translocator, such that, for instance, hexamers are strongly preferred and built up rapidly via lower n-mers. Alternatively, the aggregation may lead initially to a linear chain which is non-conducting but then folds into a ring, thereby forming an open channel, or dimers may aggregate into a double row which then splits open in the middle. These are just a few conceivable mechanisms by which the aggregation can lead directly to a defined energetically favoured channel structure without much or any contribution to the conductance from the preceding aggregation steps. In this situation there would be only one measurable conductance increment associated with the gating of the channel, but the multi-channel kinetics would be essentially unaffected.

3.8.2.2   Absolute reaction rates and preaggregation at the membrane surface

In the bilayer system the gating time constants depend on the lipid fluidity and the translocator concentration. Both observations are taken as arguments for an aggregation mechanism. One may, therefore, ask how well the observed time constants agree with admittedly rough estimates of the absolute reaction rate, assuming that the rate is diffusion limited[81].

For alamethicin the concentration in the lipid phase has been determined[82] and an approximate upper limit of the pre-exponential frequency factor, A, of the rate constants can be obtained from the number of collisions per second between the molecules. A molecule of diameter d, diffusing in the plane of the membrane with velocity v, sweeps an area per second equal to dv. At a surface concentration C, the collision frequency v encountered by one molecule is given by:

$$v = 2\,dv\,C \tag{3.11}$$

From random walk theory, the diffusion coefficient is

$$D = l^2 v'/2 \qquad (3.12)$$

where l is the length of a diffusional jump and v' the jump frequency. Combining equations (3.11) and (3.12), the collisional frequency becomes

$$v = 4D \, dC/l \qquad (3.13)$$

Neglecting steric effects, this yields the pre-exponential factor

$$A = v/C = 4Dd/l \qquad (3.14)$$

Assuming a jump length of 8 Å, which is the average distance between the two lipid molecules, a diameter of 10 Å for the width of the alamethicin and a diffusion coefficient of 10^{-8} cm^2 s^{-1}, $A = 5 \times 10^{-8}$ cm^2 s^{-1} molecule^{-1}.

At an alamethicin concentration of 10^{-5} M in the lipid phase[82], the surface concentration is 1.5×10^9 molecules cm^{-2}, resulting in collision rates of 1.12×10^{11} molecules s^{-1} cm^{-2}. The overall time constants of the conductance changes depend also on the free energies of the different aggregates and a quantitative comparison with the collision rates is, therefore, not very useful. But in the single channel experiments the transition from one conductance level to the next higher one would give a direct measure of the forward reaction rate. In these experiments the alamethicin concentration is closer to 10^7 molecules cm^{-2} and a given channel experiences, therefore, only 0.5–1 collisions s^{-1}, depending on size. The observed transitions from one conductance level to the next higher one are in most lipids 10–100 times more frequent than that[69, 70], and because the estimate neglected steric factors and activation energies which would slow the reaction even further, it seems likely that either the collision frequency is increased by a solvent cage effect keeping the colliding molecules in close vicinity or that the translocators are already preaggregated in clusters at the membrane surface. In the case of alamethicin, hydrogen bonds between the glutamine and glutamic acid residues might provide the necessary interaction points. The high viscosity of alamethicin at an air–water interface[38] and its strong aggregation tendency in aqueous solution[114, 115] also support this contention. In contrast, monazomycin has a low surface viscosity and less tendency to aggregate in water, and its time constants are generally 100 times larger than those of alamethicin.

This preaggregation may be much more pronounced with EIM or cellular translocators and may be one method by which the cell has maximised the gating rates. Effective concentrations in the range of 10^{12}–10^{13} molecules cm^{-2} can be achieved in this way and with such concentrations the calculations can easily fit the observed rates.

3.8.2.3 The voltage dependence of the aggregation rates

So far it was assumed that only the rate constants of the insertion process depend on the membrane potential, whereas the aggregation rates are unaffected by it. This need not be so because the insertion process is probably not an all-or-none phenomenon.

The average orientation and depth of insertion of the translocators can vary with the potential, especially if their interaction with the lipid hydrocarbon region is relatively high. Those molecules that reach further into the hydrocarbons may aggregate faster and as a result the aggregation rate constants may depend on the membrane potential. This additional assumption is necessary because of several experimental observations.

(1) Both in nerve[14] and in the bilayers[55] the initial maximal rate of rise of the conductance in response to a voltage step increases with the amplitude of the voltage step and does not reach a maximum at high clamp potentials. Thus there is no indication that the aggregation process becomes rate limiting when the insertion rates are higher than the aggregation rates.

(2) In viscous lipids, i.e. when the aggregation is presumably slow, there appears with alamethicin and even more so with monazomycin a rather complex sequence of early transients in response to voltage steps when these steps are applied from a holding potential at which the membrane conductance is high. During these transients the conductance changes temporarily in the opposite sense from that expected for the sign of the potential step. For example, if the step is negative, i.e. leads to a lower steady state conductance, the conductance first decreases, then increases and finally decays to the steady state value.

(3) The rate of removal of monomers from the membrane interior to the surface is limited by the dissociation rate of the aggregates and if the aggregates are energetically favoured, the dissociation rates become slow. This aspect becomes particularly obvious in the case of inactivation, where the dimers are assumed to have a low free energy. In this case the voltage dependence of the rates of recovery from inactivation does not match the observed rates, i.e. the recovery is too slow at large negative potentials (see Figure 3.15d) and the same is true for the gating currents in this situation.

(4) In the single channel experiments with alamethicin (Figure 3.12) the transition from a given conductance level to the next lower level is presumably caused by the dissociation of one monomer from the channel aggregate. The frequency of this transition is slightly voltage dependent, indicating that the corresponding rate constants also depend on the potential[110].

All these effects can be quantitatively accounted for if it is assumed that the aggregation rate constants are dependent on the membrane potential. The relation between the rate constants and the potential should be such that the rate constants vary between zero, when the translocators are at the surface, and some maximum when they are fully extended across the bilayer (see Figure 3.17). The range and effectiveness of the rate control by the voltage should also be variable. It is likely that the effect decreases with increasing size of the aggregate because their stronger interactions with the lipids would tend to keep them in a more fixed position. An empirical mathematical expression[60] satisfying these requirements assumes aggregation rate constants of the form:

$$k_{(n)(n+1)} = A \exp \{-[E_{(n)(n+1)} - W_n f(V)]/RT\} \qquad (3.15)$$

$$k_{(n+1)(n)} = A \exp \{-[E_{(n+1)(n)} + W_n f(V)]/RT\} \qquad (3.16)$$

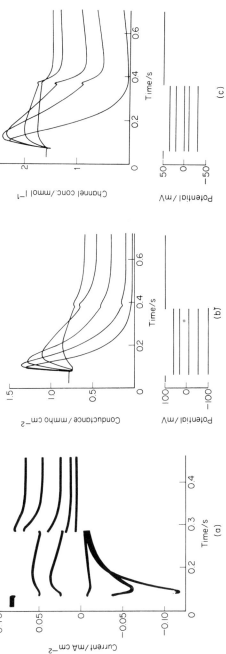

Figure 3.18 (a) Membrane currents in response to potential steps [shown in (b)] from a bilayer in 0.01 M NaCl in the presence of 10^{-6} M monazomycin on one side. The membrane solution contained 100 mg oleylphosphate and 50 mg cholesterol in 1 ml of octane. Five sweeps were superimposed on a storage scope. The potential steps were applied at 10 s intervals. The current transients start from 0.08 mA cm^{-2}, corresponding to the high conductance at the holding potential of 100 mV (positive on the side of monazomycin). Their sign inverts as the potential of the step becomes negative. Because of the slow sweep speed, the usual capacitance transients lasting only 100 μs are not visible

(b) The time course of the membrane conductance as derived from the currents in (a). Note the initial decrease followed by a rise and subsequent exponential fall. At the end of the potential step the sign of these transients inverts and their time course is slower. The transients disappear at temperatures above 40°C and are absent in less viscous lipids, e.g. glycerol diolein

(c) Channel concentration as a function of time and voltage as calculated from the model assuming that the aggregation rate constants are voltage dependent. Equations (3.15)–(3.17) were used for the rate constants, and it was assumed that only the hexamer is conductive. The following constants were used: $C = 1$, $A = 100$, $z = 2$, $E_{01} = 4$, $E_{10} = 2$, E_{12}–$E_{65} = 0$, $\varepsilon_1 = 2$, $\varepsilon_2 = 2$, $q = 0.5$, $W_1 = 2$, $W_2 = 0.5$, $W_3 = 0.5$, $W_4 = 0.4$, $W_5 = 0.4$. In the model the transient conductance increase, in response to potentials which lower the steady state conductance, results from the voltage-dependent dimer breakdown which increases the concentration of monomers temporarily, causing a transient build up of conducting hexamers from lower n-mers. The effect is dependent on a fast decay of dimers and a slower removal of inserted monomers to the surface (From Baumann and Mueller[40], by courtesy of Alan R. Liss

$$f(V) = \frac{\exp\{-[\varepsilon_1 - qV23.05]/RT\}}{\exp\{-[\varepsilon_1 - qV23.05]/RT\} + \exp\{-[\varepsilon_2 + qV23.05]/RT\}} \quad (3.17)$$

$W_n f(V)$ modifies the interaction energies between the monomers and the different oligomers as a function of the potential; ε and q control the range over which the potential acts; $f(V)$ varies between zero and one. The constants W_n are assumed to decrease with increasing oligomer size. The form of this function is empirical and chosen for convenience. A more realistic formulation would require specific and at present arbitrary assumptions about the insertion depths and aggregation rates as a function of the potential. Nevertheless, in this form the equations account very well for the entire set of kinetics seen so far in the bilayer system and nerve, including the early transients and the relation between gating currents and inactivation. An example of the early transients is shown in Figure 3.18.

3.9 SOME PROPERTIES OF THE OPEN CHANNEL

3.9.1 The instantaneous current–voltage relation

The conductance of the open channel depends on its dimensions and on the density and nature of the hydrophilic groups. It is, therefore, most likely different for the different translocators. When the channel diameter is large enough to accommodate the first hydration shell of an ion, the conductance can probably be estimated from the channel dimensions; for smaller diameters, the energies to dehydrate the ion either completely or partially may determine the rate of entrance into the channel, and the diffusion within the channel may be restricted by the spacing of the hydrophilic groups.

Measurements of the instantaneous open channel conductance for EIM[24], alamethicin[55] and monazomycin[55] as a function of the membrane potential show a hyperbolic sine relation[83], typical of a barrier diffusion mechanism. This mechanism is well understood for the unmodified membrane[84] where the ions crossing the membrane are pulled away from the membrane surface into the hydrocarbon region. Because of the multipolar nature of the surface, the field of the polar head groups falls off rather steeply and the main work to move the ion into the hydrocarbon region is done within the first 5 Å of the hydrocarbon phase. The curvature of the hyperbolic sine function is determined by this distance divided by the field. In the open channels the barrier may be formed by the repulsive potential of the gating charges, which, arranged as a ring near one channel entrance, would form a multipole of the same sign as that of the permeating ions. The hyperbolic sine curvature of the channel conductance is less than that of the unmodified membrane, indicating that the barrier is steeper, perhaps owing to the higher dielectric constant of the channel interior.

The instantaneous current–voltage curves in nerve are reported to be linear. However, reliable data have been obtained only between ± 100 mV, a range over which the bilayer curves are also almost linear. At larger potentials the gating action in nerve becomes so fast that it interferes with the measurements.

3.9.2 Ion selectivity

The ion specificity of the open channel seems to have no evident connection with the gating process. At present it is not known if the high specificities shown by most cellular systems are a property of the entire length of the channel or if they belong only to a short segment at either end of a wide and not selective portion that is responsible for the gating. A series selectivity filter could easily be accommodated by any channel model, be it of the aggregating or the preformed type.

Experiments with EIM and alamethicin have shown that large multivalent cations such as protamine or spermidine can convert the channel selectivity from cationic to anionic without changing the gating characteristics, and it is likely that the polycations act as a filter or sieve in series with the channels. Similar ideas have been proposed for nerve[85].

3.10 DOES THE AGGREGATION MECHANISM APPLY TO CELL MEMBRANES?

On the basis of the presented arguments, one may conclude that the insertion–aggregation process is a plausible mechanism of excitation for the artificial bilayer systems. An extrapolation to cell membranes rests on much shakier grounds. It is based mainly on the extensive similarities between the steady state and kinetic behaviour in both cases and perhaps to some extent on the continuity of evolution which would tend to preserve a useful principle.

Intuitively it would seem that the Na or K channels involve structural elements larger than alamethicin or monazomycin. EIM with a monomeric molecular weight near 25 000 may resemble these structures more closely, but size alone does not exclude an aggregation mechanism. Much evidence has been accumulated that membrane proteins can laterally diffuse in the plane of the bilayer[86-89] and aggregation by lateral diffusion plays a role in immunological phenomena[87, 88]. The short time constants of the gating process also do not present an obstacle, especially if one considers the possibility that only parts of the molecule are inserted into the membrane while the main body stays at the surface, perhaps serving as a selectivity filter for different ions. A rather naive and abstract model of such a larger translocator in which peptide chains are attached to a globular body is shown in Figure 3.19. The bulky part of the molecule staying at the surface would favour preaggregation so that after insertion of the side arms the molecules only have to rotate or diffuse for a very short distance, thus generating high aggregation rates.

Those aspects of the kinetics that in the bilayers support an aggregation scheme have not yet been systematically investigated in nerve. If the membrane lipid composition could be appropriately altered either by exchange or enzymatic methods, the gating rate might become slower and effects as they are seen in the bilayer in viscous lipids (Figures 3.7 and 3.18) could appear.

There is one aspect of the aggregation scheme that might decide its relevance to excitation in cells. This is the fact that the number of inserted monomers as measured from the integral of the gating currents could be considerably larger than the number of open channels. Especially if the association constants favour the lower, non-conducting, aggregates, e.g. dimers, the ratio of inserted monomers to open channels would be much larger than the number of monomers forming one channel, i.e. three in the H & H scheme or six for a hexameric channel. This would have to be especially true if the aggregation were to account quantitatively for inactivation. In this case the lowest possible theoretical ratio of inserted monomers to hexameric channels at the peak conductance is about 20:1. Unfortunately, there is no reliable data on this point for squid. The number of moving gating particles has been estimated[78] as approximately 2000 μm^{-2}, but the number of channels is not exactly known. If the estimates based on tetrodotoxin binding in lobster nerve are correct, and if the conductances of the individual Na channel are about the same in the lobster as in the squid, the Na channel density in squid as judged from the maximal Na conductance would be between 10 and 50 μm^{-2}, values that are compatible with an aggregation mechanism. On the other hand, tetrodotoxin binding studies on squid give higher channel densities but are also still of questionable reliability.

Even if the channels in nerve are opened by an aggregation mechanism, the question whether the same process also accounts for inactivation remains a separate issue. The effects of proteolytic enzymes on this process have been interpreted as supporting the existence of a separate structure that closes the channel after it was opened by the field[90], but this is not the only possible interpretation. Within the framework of an aggregation mechanism, the enzymatic action may only need to reduce the interaction energies between the monomers at the dimer level in order to account quantitatively for the observed reduction of inactivation. In this context it is also of interest to note that the reduction of the gating currents during recovery from inactivation is consistent with a process that relates the interactions between the gating particles to the inactivation, as is the case in the insertion–aggregation model. Nevertheless, the issue is wide open. There are some peculiarities

Figure 3.19 An insertion–aggregation model built from larger translocator units. The chains represent peptide loops, extending three each from a larger globular protein body (triangles). In the low conductance state, the loops rest on the membrane surface (a). An applied field pulls them into the hydrocarbon phase [(b) and (c)], where they aggregate to form hexameric (b) or trimeric (c) channels. This model is formally somewhat different from that of Figure 3.10, because the aggregation leads to an extended net of coupled channels

relating to inactivation, e.g. the dependence of inactivation on the presence of potassium ions in the aqueous phase[91, 92], which have no place in the present scheme and must await explanation until the structural details of the channels are known.

There are some interesting similarities between the electrically gated channels responsible for action potentials and channels in which ligand binding supplies the trigger for the conductance increase. The open conductance of the acetylcholine receptor channels has about the same value[93] as that of the Na channels, i.e. 10^{-10} mho, and there is an indication that the conductance is a high order function of the acetylcholine concentration. The channels also show inactivation[94, 95], the degree and rate of which depend on the ligand concentration and, as in the case of alamethicin, on the concentration of calcium in the aqueous phase. Furthermore, only a fraction of the receptor molecules seems to form open channels upon maximal activation[96] and, conversely, only a fraction of the available acetylcholine leads to the opening of channels[93]. All these observations are compatible with a mechanism in which the binding of acetylcholine leads to an aggregation of the receptor molecules or parts thereof, but the evidence is obviously not very compelling.

The proposed mechanism also invites speculation about ion pumping, especially when one considers the suggested structure of alamethicin with its ion chelating cavity (Figure 3.11). Such cavities formed by hairpin loops of hydrophobic peptide chains attached to a larger protein could move thermally back and forth through the membrane, binding and releasing the ion on alternate sides in response to changes of the ion dissociation constants caused by phosphorylation at one side and dephosphorylation at the other. A 'tethered' carrier of this type would function very much like valinomycin but, being part of a larger molecular complex, it would be locally restricted and its operation could be controlled by other parts of the complex.

3.11 ALTERNATIVE MODELS

Finally, one may ask if the H & H equations are applicable to alternative mechanisms that do not involve the linear aggregation of freely diffusing channel precursors. The first term of the reaction scheme (3.1) requires the movement of charged particles in the field and can, therefore, only be realised by something moving normal to the membrane plane and through a major portion of the resistive barrier. This requirement could either be met by peptide chains moving through the hydrocarbon region, as was postulated for the insertion of alamethicin, or by the intramolecular movement of parts of the channel structure. Formally, both cases would be identical. The second part of the scheme, $g \approx (P_1)^n$, however, does not need to be interpreted as describing a polymerisation or aggregation reaction. The most direct interpretation—also mentioned by H & H—is a statistical one, i.e. n particles must occupy simultaneously a particular position within the membrane in order for the channel to open, but they do not have to interact with each other. One possible realisation of this mechanism, involving the removal of six blocking groups by the field, is shown in Figure 3.20.

However, in order to account for the observed state dependence of the conductance and of the gating currents, the structural elements carrying the gating charge would have to interact with each other, and thus the second part of the H & H equation must again be modified into a more explicit kinetic scheme. When these interactions are taken into account, the model is formally and in its molecular aspects only a modification of the aggregation model in the sense that the aggregation of the gating structure takes place in a restricted space and between components of a larger molecular assembly instead of between freely diffusing entities, and it is conceivable that this type of structure evolved from the earlier and more primitive aggregation mechanism. Quantitative calculations on this model are not yet available and so its potential as a viable alternative to the aggregation process is unknown.

Other kinetic models that assume a series of sequential, voltage induced configurational changes leading to the opening of a gate and subsequent inactivation[97, 98] are also consistent with the H & H scheme, but their interpretation in molecular terms has not yet been attempted.

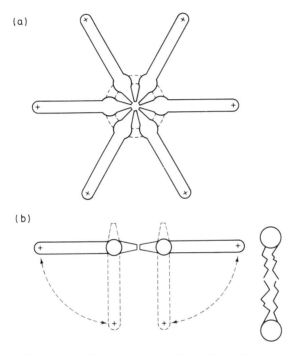

Figure 3.20 Schematic model of an electrically gated channel; (a) top view, (b) side view. In this case the gating action does not involve the aggregation of freely diffusing monomers. Instead, the opening of the channel requires that all six side arms tilt normal to the membrane plane, as seen in (b) (dashed lines). This removes the six groups that block the channel entrance in the closed state. Such a structure could be formed from a single peptide chain as indicated in (a) (solid line) or by the aggregation of six monomers shaped similar to alamethicin. Below the gating structures, extending through the membrane, there may be a fixed open channel, indicated by the dashed line in (a). Alternatively, the side arms themselves, when tilted into the membrane, may form a 'barrel stave' channel. This figure and Figure 3.19 are intended to show how the primitive insertion–aggregation mechanism shown in Figure 3.10 might have evolved into a more complex system while maintaining the same basic principles of operation

Having placed so much emphasis on one particular molecular model that is consistent with the original H & H formulation and can account for the bulk of the experimental data, it is only appropriate to point out that this obvious bias is not intended to exclude or ignore the many alternatives that have been suggested but were not mentioned here, e.g. Refs. 98–105. As more chemical and kinetic data emerge, the possible choices among these models become increasingly restricted, and, even now, any scheme worth considering must be consistent with much more than the minimal H & H kinetics. For the cellular systems the problem can only be solved after the translocator structures are known and only then shall we know if the mechanism of membrane excitability is essentially the same in all cells.

Acknowledgements

The reviewer would like to thank G. Baumann and D. O. Rudin for many discussions and suggestions, and H. Freifelder, S. Montimore and C. Williams for help with the preparation of the manuscript. The work was supported by a grant from the National Science Foundation (GB 8144-01).

References

1. Osterhout, W. J. and Hill, S. E. (1939). *J. Gen. Physiol.*, **22**, 115
2. Blinks, L. R. (1936). *J. Gen. Physiol.*, **19**, 633
3. Pickard, B. G. (1974). *Naturwiss.*, **61**, 60
4. Eckert, R. (1972). *Science*, **176**, 473
5. Maeno, T. (1959). *J. Gen. Physiol.*, **43**, 139
6. Yamamoto, T. (1961). *Int. Rev. Cytol.*, **12**, 361
7. Finkelstein, A. (1964). *J. Gen. Physiol.*, **47**, 545
8. Hille, B. (1970). *Progr. Biophys. Mol. Biol.*, **21**, 1
9. Hagiwara, S. and Naka, K. (1965). *J. Gen. Physiol.*, **48**, 141
10. Grundfest, H. (1971). In *Biophysics and Physiology of Excitable Membranes*, 477 (W. Adelman, editor) (New York: Van Nostrand Reinhold)
11. Inoue, I., Ishida, N. and Kobatake, K. (1973). *Biochim. Biophys. Acta*, **330**, 27
12. Hodgkin, A. L., Huxley, A. F. and Katz, B. (1952). *J. Physiol. (London)*, **116**, 424
13. Hodgkin, A. L. and Huxley, A. F. (1952). *J. Physiol. (London)*, **116**, 449
14. Hodgkin, A. L. and Huxley, A. F. (1952). *J. Physiol. (London)*, **116**, 473
15. Hodgkin, A. L. and Huxley, A. F. (1952). *J. Physiol. (London)*, **116**, 497
16. Hodgkin, A. L. and Huxley, A. F. (1952). *J. Physiol. (London)*, **117**, 500
17. Mueller, P. and Rudin, D. O. (1968). *J. Theoret. Biol.*, **18**, 222
18. Mueller, P. and Rudin, D. O. (1969). In *Current Topics in Bioenergetics*, Vol. 3, 157 (D. R. Sanadi, editor) (New York: Academic Press)
19. Montal, M. and Mueller, P. (1972). *Proc. Nat. Acad. Sci. USA*, **69**, 3561
20. Bangham, A. D., Standish, M. M. and Watkins, J. C. (1965). *J. Mol. Biol.*, **13**, 238
21. Razin, S., Morowitz, H. J. and Terry, T. M. (1965). *Proc. Nat. Acad. Sci. USA*, **54**, 219
22. Kagawa, Y. and Racker, E. (1971). *J. Biol. Chem.*, **246**, 5477
23. Reeves, J. P. and Dowben, R. M. (1969). *J. Cell. Physiol.*, **73**, 49
24. Mueller, P. and Rudin, D. O. (1963). *J. Theoret. Biol.*, **4**, 268
25. Mueller, P. and Rudin, D. O. (1968). *Nature*, **217**, 713
26. Hubbell, W. L. and McConnell, H. M. (1968). *Proc. Nat. Acad. Sci. USA*, **61**, 12
27. Keith, A. D., Sharnoff, M. and Cohn, G. E. (1973). *Biochim. Biophys. Acta*, **300**, 379
28. Chapman, D. (1966). *Ann. N.Y. Acad. Sci.*, **137**, 745
29. Elson, E. L. and Magde, D. (1974). *Biopolymers*, **13**, 1

30. Magde, D., Elson, E. L. and Webb, W. W. (1974). *Biopolymers*, **13**, 29
31. Shinitzky, M., Dianoux, A. C., Gitler, C. and Weber, G. (1971). *Biochemistry*, **10**, 2106
32. Steim, J. M., Tourtellotte, M. E., Reinert, J. C., McElhaney, R. N. and Rader, R. L. (1969). *Proc. Nat. Acad. Sci. USA*, **63**, 104
33. Levine, Y. K. and Wilkins, M. H. F. (1971). *Nature New Biol.*, **230**, 69
34. Cain, J., Santillan, G. and Blasie, J. K. (1972). In *Membrane Research, Proc. 1972 ICN-UCLA Symp. Mol. Biol.* (C. F. Fox, editor) (New York: Academic Press)
35. Engelman, D. (1970). *J. Mol. Biol.*, **47**, 115
36. Wilkins, M. H. F., Blaurock, A. E. and Engelman, D. M. (1971). *Nature New Biol.*, **230**, 72
37. Griffith, O. H., Dehlinger, P. J. and Van, S. P. (1974). *J. Membrane Biol.*, **15**, 159
38. Mueller, P. (1974). Unpublished observations
39. Meyer, C. E. and Reusser, F. (1967). *Experientia*, **23**, 85
40. Mitscher, L. A., Shay, A. J. and Bohonos, N. (1967). *Appl. Microbiol.*, **15**, 1002
41. Bohlmann, F., Dehmlow, E. V., Neuhahn, H. J., Brandt, R. and Bethke, H. (1970). *Tetrahedron*, **26**, 2199
42. Payne, J. W., Jakes, R. and Hartley, B. S. (1970). *Biochem. J.*, **117**, 757
43. Chandler, W. K., Hodgkin, A. L. and Meves, H. J. (1965). *J. Physiol. (London)*, **180**, 821
44. Gilbert, D. L. (1971). In *Biophysics and Physiology of Excitable Membranes*, 359 (W. J. Adelman, editor) (New York: Van Nostrand Reinhold)
45. Goldman, D. E. and Blaustein, M. P. (1966). *Ann. N.Y. Acad. Sci.*, **137**, 967
46. McLaughlin, S. G. A., Szabo, G., Eisenman, G. and Ciani, S. M. (1970). *Proc. Nat. Acad. Sci. USA*, **67**, 1268
47. Hodgkin, A. L. and Chandler, W. K. (1965). *J. Gen. Physiol.*, **48**, part 2, 27
48. Muller, R. U. and Finkelstein, A. (1972). *J. Gen. Physiol.*, **60**, 285
49. Cherry, R. J., Chapman, D. and Graham, D. E. (1972). *J. Membrane Biol.*, **7**, 325
50. Adrian, R. H., Chandler, W. K. and Hodgkin, A. L. (1970). *J. Physiol. (London)*, **208**, 607
51. Luttgau, H. C. (1960). *Arch. Ges. Physiol.*, **271**, 613
52. Ehrenstein, G. and Gilbert, D. L. (1966). *Biophys. J.*, **6**, 553
53. Schwartz, J. R. and Vogel, W. (1971). *Experientia*, **27**, 397
54. Grundfest, H. (1965). *J. Gen. Physiol.*, **49**, 321
55. Mueller, P. (1975). In preparation
56. Goldman, L. and Schauf, C. L. (1973). *J. Gen. Physiol.*, **61**, 361
57. Armstrong, C. M. and Bezanilla, F. (1973). *Nature*, **242**, 459
58. Bezanilla, F. and Armstrong, C. M. (1974). *Science*, **183**, 753
59. Keynes, R. D. and Rojas, E. (1973). *J. Physiol. (London)*, **233**, 28P
60. Baumann, G. and Mueller, P. (1974). *J. Supramol. Struct.*, **2**, 516
61. Finkelstein, A. and Holz, R. (1972). In *Membranes*, Vol. 2, 377 (G. Eisenman, editor) (New York: Marcel Dekker)
62. Glaser, M., Simpkins, H., Singer, S. J., Sheetz, M. and Chan, S. I. (1970). *Proc. Nat. Acad. Sci. USA*, **65**, 721
63. Moran, A. and Ilani, A. (1974). *J. Membrane Biol.*, **16**, 237
64. Goodall, M. C. (1970). *Biochim. Biophys. Acta*, **219**, 28
65. Goodall, M. C. (1971). *Arch. Biochem. Biophys.*, **147**, 1971
66. Bukovsky, J. (1974). Personal communication
67. Ehrenstein, G., Lecar, H. and Nossal, R. (1970). *J. Gen. Physiol.*, **55**, 119
68. Bean, R. C., Shepherd, W. C., Chan, H. and Eichner, J. (1969). *J. Gen. Physiol.*, **53**, 741
69. Gordon, L. G. M. and Haydon, D. A. (1972). *Biochim. Biophys. Acta*, **255**, 1014
70. Eisenberg, M., Hall, J. E. and Mead, C. A. (1973). *J. Membrane Biol.*, **14**, 143
71. Hladky, S. B. and Haydon, D. A. (1970). *Nature*, **225**, 451
72. Baumann, G. and Mueller, P. (1975). In preparation
73. Hoyt, R. C. (1971). *Biophys. J.*, **11**, 110
74. Hoyt, R. C. and Adelman, W. J. (1970). *Biophys. J.*, **10**, 610
75. Peller, L. and Barnett, L. (1962). *J. Phys. Chem.*, **66**, 680
76. Scheele, R. B. and Schuster, T. M. (1974). *Biopolymers*, **13**, 275
77. Cantor, C. R. (1968). *Biopolymers*, **6**, 369
78. Keynes, R. D. (1974). Personal communication
79. Armstrong, C. (1974). Personal communication

80. Meissner, H. P. (1965). *Arch. Ges. Physiol.*, **283**, 213
81. Smoluchowski, M. V. (1917). *Z. Physiol. Chem.*, **92**, 129
82. Chelack, W. S. and Petkau, A. (1973). *J. Lipid Res.*, **14**, 255
83. Frenkel, J. (1955). *Kinetic Theory of Liquids*, 40 (New York: Dover Publications)
84. Kauffman, J. W. and Mead, C. A. (1970). *Biophys. J.*, **10**, 1084
85. Hille, B. (1974). In *Membranes*, Vol. 3 (G. Eisenman, editor) (New York: Marcel Dekker)
86. Blasie, J. K., Worthington, C. R. and Dewey, M. M. (1969). *J. Mol. Biol.*, **39**, 407
87. Matus, A., de Petris, S. and Raff, M. C. (1973). *Nature New Biol.*, **244**, 278
88. Frye, L. D. and Edidin, M. (1970). *J. Cell. Sci.*, **7**, 319
89. Cone, R. A. (1972). *Nature New Biol.*, **236**, 39
90. Armstrong, C. M., Bezanilla, F. and Rojas, E. (1973). *J. Gen. Physiol.*, **62**, 375
91. Adelman, W. J. and Palti, Y. (1969). *J. Gen. Physiol.*, **54**, 589
92. Chandler, W. K. and Meves, H. (1970). *J. Physiol. (London)*, **211**, 623
93. Katz, B. and Miledi, R. (1972). *J. Physiol. (London)*, **224**, 665
94. Manthey, A. A. (1972). *J. Membrane Biol.*, **9**, 319
95. Nastuk, W. L., Manthey, A. A. and Gissen, A. J. (1966). *Ann. N.Y. Acad. Sci.*, **137**, 999
96. Miledi, R. and Potter, L. T. (1971). *Nature*, **233**, 599
97. Armstrong, C. M. (1969). *J. Gen. Physiol.*, **54**, 553
98. Moore, J. W. and Cox, E. B. (1974). Personal communication
99. Goldman, D. E. (1964). *Biophys. J.*, **4**, 167
100. Goldman, D. E. (1971). In *Biophysics and Physiology of Excitable Membranes*, 357, (W. J. Adelman, editor) (New York: Van Nostrand Reinhold)
101. Changeux, J. P., Thiery, J., Tung, Y. and Kittel, C. (1967). *Proc. Nat. Acad. Sci. USA*, **57**, 335
102. Hill, T. L. and Chen, Y. D. (1972). *Biophys. J.*, **12**, 960
103. Adam, G. (1968). *Z. Naturforsch.*, **23b**, 181
104. Bretag, A. H., Davis, B. R. and Kerr, D. I. B. (1974). *J. Membrane Biol.*, **16**, 363
105. Offner, F. F. (1972). *Biophys. J.*, **12**, 1583
106. Miller, C. and Mueller, P. (1974). Unpublished observations
107. Gilbert, D. L. and Ehrenstein, G. (1969). *Biophys. J.*, **9**, 447
108. Cole, K. S. and Moore, J. W. (1960). *Biophys. J.*, **1**, 1
109. Silverman, D. N. and Scheraga, H. A. (1972). *Arch. Biochem. Biophys.*, **153**, 449
110. Eisenberg, M. (1974). Personal communication
111. Hille, B. (1968). *J. Gen. Physiol.*, **51**, 221
112. Frankenhaeuser, B. and Hodgkin, A. L. (1957). *J. Physiol. (London)*, **137**, 218
113. Heyer, E. J. (1974). Personal communication
114. Lau, L. Y. and Chan, S. I. (1974). Personal communication
115. McMullen, A. I. and Stirrup, J. A. (1971). *Biochim. Biophys. Acta*, **241**, 807

4
Energy Transducing Mechanisms in Muscle

Y. TONOMURA and A. INOUE
Osaka University

Abbreviations

εATP $1,N^6$-ethenoadenosine triphosphate
CMB   *p*-chloromercuribenzoate
NEM   *N*-ethylmaleimide

PEP   phosphoenolpyruvate
SH-ATP 6-mercaptopurine ribose triphosphate
TCA   trichloroacetic acid

4.1 INTRODUCTION

Muscle transduces chemical energy to mechanical work. Muscle contraction was one of the first types of energy transduction to be studied, since it can be easily observed and the accompanying change is large. A unique character of muscle contraction is its control, since the fastest muscle can attain maximum tension from rest within a few milliseconds. The structural basis of these characteristics of muscle contraction is that muscle consists of myofibrils, composed of two kinds of filaments made of different proteins, myosin and actin, and that contraction is caused by the interaction between these two filaments.

ATP and Ca^{2+} ions are low molecular weight substances, which react with these two protein filaments during muscle contraction. Biochemical studies on metabolism in muscle suggested that the chemical reaction most likely to provide the energy for contraction is hydrolysis of ATP. Engelhardt[1], Szent-Györgyi[2], Weber[3] and others showed that myosin has ATPase activity[1], which is activated markedly by actin in the presence of Mg^{2+} ions[2], and that glycerol-treated muscle fibres[3], myofibrils[4] and actomyosin threads[5,6] contract on adding ATP, with concomitant hydrolysis of ATP. Therefore, it has generally been accepted that ATP is the energy source of muscle contraction. Using muscle in which creatine kinase, glycolysis and respiration were all inhibited, Cain and Davies[7] showed more recently that the amount of ATP decreases during contraction and that the decrease in ATP corresponds to the energy consumption of the muscle.

A. F. Huxley and Niedergerke[8] and Hanson and H. E. Huxley[9] found that the length of the *A* band remains constant when muscle fibres are extended or contracted, and that the length of the *H*-zone changes by the same amount as the alteration in the length of the sarcomere. Based on these results, it was predicted that there must be two kinds of filaments in myofibrils, both having a fixed length. Contraction must therefore occur by means of a change in their positions owing to sliding of the filaments. This 'sliding theory' is supported by various lines of evidence. It was found by electron microscopy[10,11] that the thick and the thin filaments are regularly arranged in myofibrils, and the mutual positions of these two filaments change when muscle fibres extend or contract, just as the sliding theory predicts. Furthermore, it was discovered that the thick filaments are mainly composed of myosin and the thin filaments mainly of actin[12-15], and that the development of tension of living muscle[16,17] and the ATPase activity of living muscle[18] and glycerol-treated muscle fibres[19,20] are both proportional to the area of the overlapping region of these two filaments. Moreover, we[21] showed that single sarcomeres, which are composed only of the thick and the thin filaments, contract on adding ATP. The fine structure of the sites where tension is developed and ATP is hydrolysed was investigated by analysis of x-ray diffraction patterns[22], electron microscopy[23], and biochemical studies on the structural proteins[24], i.e. myosin, actin and others. From these investigations it was proposed that *the head portions of myosin molecules form projections from the thick filaments*, that *cross-bridges are formed by the binding of the projections with the thin filaments*, and that *sliding of the thin filaments past the thick filaments occurs as a result of the movement of cross-bridges, which is coupled with ATP-splitting.*

Heilbrunn[25,26] and Kamada and Kinoshita[27] found that muscle contraction is induced when Ca^{2+} ions are injected into the muscle fibre. Later, it was shown that glycerol-treated muscle fibres contract on adding Mg^{2+}–ATP in the presence of a small amount of Ca^{2+} ions and relax on removal of Ca^{2+} ions with a chelate compound[28,29]. Hasselbach and Makinose[30] and Ebashi and Lipmann[31] discovered that Ca^{2+} ions are actively transported in the sarcoplasmic reticulum, coupled with the ATPase reaction. The mechanism of this Ca^{2+} uptake will be discussed in Section 4.5.2. Ebashi[32] discovered that relaxation with removal of Ca^{2+} ions requires the presence of tropomyosin and troponin located on the thin filaments and that troponin is the Ca^{2+}-receptor protein in muscle.

Since several excellent reviews[32,33] have already been published on the mechanism of control of muscle contraction by Ca^{2+} ions, this review will mainly deal with recent biochemical research on the molecular mechanism of coupling of the chemical reaction of ATP hydrolysis with production of mechanical work. The sliding theory gives us a framework to consider the molecular mechanism of muscle contraction. However, the sliding theory originally predicted only structural changes in muscle observed before and after contraction. In other words, the sliding theory was presented in only one sense without elucidating the molecular mechanism for the structural changes during muscle contraction. According to the sliding theory, there are at least three fundamental steps in muscle contraction: (i) the attachment of the projections from the thick filaments to the thin filaments, (ii) movement of the cross-bridges thus formed, which induces sliding of the thin filaments past the thick filaments, and (iii) the detachment of the projections from the thin filaments. Accordingly, studies on the molecular mechanisms of these three fundamental steps have been made from three different directions. One is by study of the fine structure of the contractile apparatus, especially the movement of projections from the thick filaments, using x-ray diffraction, electron microscopy and fluorescence polarisation. The second is by physiological studies in the dynamic characteristics of muscle to elucidate the kinetic movement of projections at the molecular level. The third is by biochemical studies on the reactions between myosin, actin and ATP.

The first and second types of experiments were mainly used in early studies on the molecular mechanism of muscle contraction, while the third, which deals with the ATPase reaction, has only recently received attention. However, this method has been found useful for clarification of the molecular mechanism, and has led to discovery of many intermediates of the myosin–ATPase reaction, such as the reactive myosin–phosphate–ADP complex. Therefore, studies on the mechanism of the ATPase reaction have attracted many research workers as one of the most effective ways to clarify the molecular mechanism of muscle contraction, and there have been frequent attempts to explain the results of structural and physiological experiments on the basis of the reaction mechanism of the myosin–actin–ATP system.

This review is mainly on the reaction mechanism of the actin–myosin–ATP system, and the molecular mechanism of contraction is discussed from the biochemical point of view. We have already reported on the reaction mechanisms of the myosin– and the actomyosin–ATPase in a monograph[34] and in review articles[35, 36]. However, there is still diversity of opinion in this field, and there have been many important discoveries since these publications. Therefore, in this review facts and theories on the molecular mechanism of the ATPase reaction are described in detail, with emphasis on recent developments.

4.2 SUBSTRUCTURE OF THE MYOSIN MOLECULE

As described in the previous section, it is now well established that the splitting of ATP and development of tension both result from the interaction between the thin filaments, which are mainly composed of F-actin, and the projections from the thick filaments, which are mainly composed of myosin. Biochemical studies on the actin–myosin–ATP system have shown that *a reaction intermediate, the reactive myosin–phosphate–ADP complex, is formed by the reaction of myosin with ATP and its decomposition is accelerated by F-actin*, and that *active movement of the heads of myosin molecules, which is coupled with decomposition of the intermediate, induces passive sliding of the thin filaments past the thick filaments*[34]. This section describes the substructure of the myosin molecule which forms the thick filaments of myofibrils, and the part of the molecule in which the binding site with actin and the active site of ATPase are located. There have been many reviews[34, 36, 37] on the substructure of myosin, so the subunit structure of myosin will be mentioned here only briefly, in relation to the reaction mechanisms of myosin– and actomyosin–ATPases, which are the main topics of this review. Furthermore, it should be noted that the substructure of myosin depends on its origin, e.g. skeletal, cardiac or smooth muscle. Here the substructure of myosin from rabbit white skeletal muscle is described, since this myosin has been used most by workers in this field.

Physicochemical studies on myosin in solution and electron microscopic studies on myosin molecules showed[34] that myosin has a molecular weight of 4.6–4.8×10^5, with a total length of about 1600 Å. Lowey *et al.*[24, 38], in electron microscopic studies on myosin and its subfragments using a rotatory shadowing technique, showed that the myosin molecule has two separate heads, each with a diameter of 90 Å. Biochemical studies on subfragments of myosin revealed that both the binding site with actin and the active site of ATPase are located in the head portions which compose the projections from the thick filaments, while the tail portions aggregate at low ionic strength and constitute the back-bone of the thick filaments.

A scheme of the myosin molecule and the structures of its various subunits are shown in Figure 4.1. One mole of heavy meromyosin, HMM, with a molecular weight of 3.4×10^5 and 1 mole of light meromyosin, LMM, with

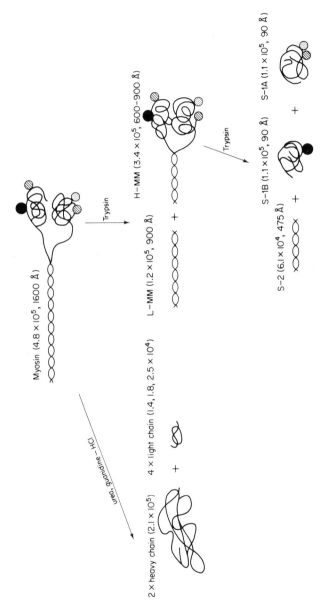

Figure 4.1 Relationship of the myosin molecule to its subunits. Values in parentheses indicate molecular weight and length. Changes in the light chains during tryptic digestion are depicted according to Hayashi's results. ●, ○ and ○ indicate g₁, g₂ and g₃, respectively. (From Tonomura³⁴, by courtesy of the University of Tokyo Press.)

a molecular weight of 1.2×10^5 are produced by tryptic digestion of 1 mole of myosin[39]. HMM is derived from the head portions and has ATPase activity and ability to bind with actin, while LMM is derived from the tails of myosin molecules and aggregates at low ionic strength. Further tryptic digestion of 1 mole of HMM yields 2 moles of subfragment-1, S-1, with a molecular weight[40] of about 11.0×10^4 and 1 mole of subfragment-2, S-2, with a molecular weight[41] of about 6×10^4. S-1 is derived from the head parts of myosin molecules, and has both the active site of ATPase and the binding site for actin. S-2 is considered to function as the hinge which connects the two heads and the tail in the myosin molecule. Furthermore, fluorescent polarisation decay studies on myosin and its subfragments[42] have revealed high bending flexibility in the connections between the heads and S-2, and S-2 and the tail of the myosin molecule.

Thus, the myosin molecule has two heads, and the heads split ATP and combine with F-actin. The problem of whether the two heads are identical or not is one of the most important ones for elucidation of the molecular mechanism of muscle contraction. Our studies on the substructure of myosin, chemical modifications of the active sites and bindings of ATP and its analogues to myosin have provided strong evidence that the structures and functions of the two heads of the myosin molecule are different from each other, as described later. However, although there has been much work on this problem, it is very difficult to obtain any conclusive evidence[34, 36], since it is uncertain whether myosin preparations are homogeneous or not. In fact, Starr and Offer[43] suggested from analysis of the N-terminal amino acid sequence of myosin that myosin is composed of at least two isoenzymes, with slightly different chemical structures. In this article the substructure of myosin is discussed, assuming tentatively that these isoenzymes have the same kinetic properties of ATPase activity.

Tsao[44] first reported that on treatment with urea, myosin dissociates into several polypeptide chains. Subsequently, it has been established by many workers[45-48] that myosin consists of two heavy chains, f, with a molecular weight of about 2×10^5 and light chains with molecular weights of about 2.5, 1.8 and 1.4×10^4. We will call these three kinds of light chains g_1, g_2 and g_3, respectively, in decreasing order of molecular weight. The fact[43, 49, 50] that HMM and S-1 both contain all the N-terminal amino acids of myosin indicates that the location of the heavy chains in the myosin molecule has polarity. The molar ratios of the three light chains, g_1, g_2 and g_3, in the myosin molecule were estimated[51, 52] to be about $1:2:1$.

The function of the light chains has attracted much attention, since Stracher[53] and Dreizen and Gershman[54] reported that when separated, the heavy and light chains have no ATPase activity, but that myosin reconstituted from the separated heavy and light chains has the ATPase activity of the original molecule. However, reconstitution of myosin from the isolated chains is very difficult, and has not yet been confirmed by others. Some chemical modifiers which inactivate myosin–ATPase have been shown to bind to the heavy chains[34]. Therefore, it is obvious that the heavy chains are essential for ATPase activity, while the function of the light chains remains to be clarified.

Using a fluorescence labelled antibody technique, Lowey and Risby[55] found that the light chains are located in the projections of the thick filaments, i.e. the head portions of myosin. The result that 1 mole of myosin has 1 mole of g_1 and g_3 apparently suggests the heterogeneity of structure of the two heads of the myosin molecule. Stracher[53] and Lowey and Weeds[51] first supposed that g_3 is derived from g_1, but Weeds and Franc[56] have recently shown that although the chemical structures of the overlapping parts of g_1 and g_3 are very similar, they differ in several residues. We[57] showed that only 1 of the 2 moles of g_2 or its derivative is removed from 1 mole of myosin or HMM by treatment with p-chloromercuribenzoate (CMB), although the two g_2 have the same amino acid composition. Moreover, 2 moles of S-1 contain only 1 mole of g_2 derivative, since only one g_2 derivative in HMM, which was removed by CMB treatment, was easily digested with trypsin. Hayashi[52] followed the change in the subunit structure of myosin during tryptic digestion using SDS gel electrophoresis. Tentatively assuming that the molar ratio of derivatives of g_1, g_2 and g_3 in S-1 was $1:1:1$, he concluded that all the results can be interpreted assuming that S-1 contains two kinds of subunits, S-1A and S-1B, in an equal molar ratio (Figure 4.1). S-1A, which has no ATPase activity, consists of four polypeptide chains with molecular weights of 5.2–5.5 (f′) and 2.7 (f″), 1.6 (g_2') and 1.4×10^4 (g_3). The former two chains are derived from the heavy chains and the latter two from g_2 and g_3 itself, respectively. S-1B, which has ATPase activity, is composed of f′, f″ and a light chain with a molecular weight of 2.1×10^4 (g_1'') derived from g_1. In the experiments the origin of the polypeptide chains in S-1 was followed only by determining the patterns of change in molecular weights. Therefore, these tentative conclusions must be confirmed by determining the chemical structure of each peptide chain.

Studies on the function of myosin also support the heterogeneity of structure and function of the heads in the myosin molecule. We[58] measured the binding of myosin with inorganic pyrophosphate (PP_i), which is a competitive inhibitor of myosin–ATPase, using an equilibrium dialysis method, and showed that 2 moles of PP_i bind to 1 mole of myosin. This result was supported by other workers[59], and a similar result was also reported for the binding of ADP to myosin[60]. Furthermore, our studies[58] suggested that 2 moles of PP_i bind to 1 mole of myosin with different dissociation constants, and that only 1 mole of PP_i binds to 1 mole of myosin constituent in actomyosin, and that this binding of PP_i causes the dissociation of actomyosin. Morita et al.[61, 62] have recently shown that in the presence of Mg^{2+} ions at low ionic strength, and especially in the presence of Mn^{2+} ions, the binding constants of 2 moles of ADP with 1 mole of HMM are very different, and the one which binds more strongly induces the change in the u.v. absorption of HMM. The heterogeneity of function of the two heads is also suggested by chemical modifications of the active sites in the myosin molecule (see Ref. 34 for details of these studies).

Thus, the relationship between the structure and function of the myosin molecule seems to be elucidated fairly well from studies on the substructure of myosin and its bindings to ADP and PP_i. However, it is very difficult to decide conclusively whether the two heads of the myosin molecule are identical or not, since it is uncertain whether myosin preparations are homogeneous or not, as described above. More conclusive evidence on this problem was obtained by kinetic studies on the myosin– and actomyosin–ATPase reactions, as described in the next two sections.

4.3   REACTION MECHANISM OF MYOSIN–ATPase

Elucidation of the reaction mechanism of myosin–ATPase is most important for clarification of the mechanism of muscle contraction in molecular terms. Kinetic methods are very valuable and often most effective for elucidation of the mechanism of an enzyme reaction. However, in many cases the simplest mechanism which can account quantitatively for the results obtained under limited experimental conditions has been proposed, and accepted as correct. We think that for an enzyme reaction which has important physiological functions and a very complicated mechanism, such as that of myosin–ATPase, kinetic studies must be pursued from various standpoints. Thus, the rate must be measured under as wide a range of experimental conditions as possible, with special regard to physiological functions of the enzyme, and the elementary steps which constitute the enzyme reaction must be measured separately. Furthermore, when several different reaction mechanisms have been proposed, they must be examined by experiments which can clearly distinguish between them.

From kinetic studies based on these principles we concluded that the reaction mechanism of myosin–ATPase is very complex but that each elementary step which constitutes myosin–ATPase has its own physiological function. In this review the kinetics of the ATPase reaction in the steady state are described first, and then the transient kinetics of the ATPase reaction are discussed. It is well known that the reaction of myosin–ATPase is dependent on the presence and absence of cations, especially divalent cations. In this article the results obtained in the presence of several mM Mg^{2+} ions are mainly analysed, since muscle fibres contain Mg^{2+} ions at a level of about 13 mM.

4.3.1   ATPase reaction in the steady state

It was first shown by Ouellet et al.[63] and by us[64] that the dependence of the rate of ATPase reaction in the steady state, v_0, on the ATP concentration follows a Michaelis–Menten equation at least at ATP concentrations above several µM. Thus, it was concluded that there are at least two steps in the reaction:

$$M + ATP \rightleftharpoons [\text{reaction intermediates}] \rightarrow M + ADP + P_i$$

In this scheme M denotes the active centre of myosin–ATPase. The dependence of v_0 on ATP concentration was recently remeasured by Lymn and Taylor[65] and by us[66] over wide ranges of ATP concentrations. It was shown that the Michaelis constant, K_m, and the maximum velocity, V_{max}, differ in different ranges of ATP concentrations. For example, in 0.5 M KCl and at pH 7.8 and 0 °C the K_m, and V_{max} in the range of ATP concentrations above 0.5 µM were 1 µM and 0.44 min^{-1}, respectively, while in thé range of ATP concentrations below 0.5 µM, V_{max} was about one quarter of that at high ATP concentrations and K_m was too low to measure accurately ($\lesssim 0.1$ µM), as shown in the inserted figure of Figure 4.2.

One of the most useful methods to clarify the mechanism of this dependence of v_0 on ATP concentration was measurement of the amount of nucleotide bound to myosin during the ATPase reaction. This was previously done by Bowen and Evans[67] and Schliselfeld and Bárány[68]. They reported the binding of about 2 moles of nucleotides per mole of myosin, but they could not determine whether ATP or ADP was bound to myosin. Therefore, we[69] measured the amounts of ATP and ADP bound to myosin during the ATPase reaction by the following methods. An ATP-regenerating system (pyruvate kinase and phosphoenol pyruvate, PEP) was coupled with the myosin–ATPase reaction to reconvert the free ADP produced to ATP very quickly. The amount of bound ADP was measured as the amount of ADP, determined by separating nucleotides using polyethyleneimine–cellulose thin layer chromatography after stopping the reaction with trichloroacetic acid (TCA). The amount of total nucleotides bound to myosin was determined from the difference between rates of diffusion of nucleotides through a membrane in the presence and absence of myosin, using a rapid-flow dialysis method. Then the amount of bound ATP was determined by subtracting the amount of bound ADP from that of total bound nucleotides:

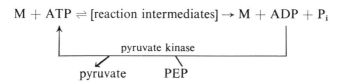

As shown in Figure 4.2, when the amount of ATP added was less than about 0.6 mole per mole of myosin, almost all the nucleotide bound strongly to myosin was ADP, and the binding of ATP to myosin was observed only when the amount of ATP added was more than about 0.6 mole per mole of myosin. The maximum amount of bound ATP was 1 mole per mole of myosin, and the dissociation constant of the binding was equal to the K_m value of myosin–ATPase in the steady state at high concentrations of ATP. These results are easily explained as follows. At ATP concentrations above 0.5 µM the Michaelis complex of the ATPase reaction is the myosin–ATP complex, while at ATP concentrations below 0.5 µM it is the complex of myosin with ADP. As described later, the transient kinetics of the ATPase reaction showed that the complex of myosin and ADP also contains bound phosphate, and we called this the reactive myosin–phosphate–ADP complex, M_P^{ADP}. The maximum amount of bound ADP depended on the KCl concentration. It was 1 mole per mole of myosin at high KCl concentrations, but decreased to less than the stoichiometric amount with decrease in the KCl concentration. For example, in 0.125 M KCl at 20°C it was 0.4 mole per mole of myosin. This result was explained by supposing that the decomposition of M_P^{ADP} involves at least two steps, since the rate of M_P^{ADP} formation under the conditions used was much higher than that of its decomposition:

$$M + ATP \overset{\text{very fast}}{\rightleftharpoons} M_P^{ADP} \rightarrow {}^{\circ}M + ADP + P_i \rightarrow M + ADP + P_i$$

130

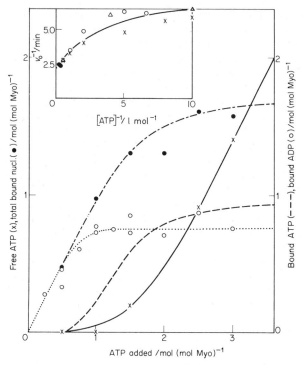

Figure 4.2 Binding of ADP and ATP to myosin during the ATPase reaction. 3.6 mg ml^{-1} myosin, 4 mg ml^{-1} pyruvate kinase, 1 mM PEP, 0.5 M KCl, 2 mM MgCl$_2$, 50 mM Tris-HCl, pH 7.8, 0 °C. The amounts of free ATP ($\times$) and total bound nucleotides (●) were measured by a rapid-flow dialysis method. The amount of ATP bound to myosin (— — —) was calculated from the concentration of free ATP measured (—$\times$—), by assuming that 1 mole of ATP binds to 1 mole of myosin with a dissociation constant of 1 μM. The amount of ADP bound to myosin (○) was measured after stopping the reaction with TCA and separating nucleotides by thin layer chromatography. The amount of total bound nucleotides (—·—·—) obtained as the sum of the amount of bound ADP measured and that of bound ATP, as calculated above, was in good agreement with the amount of total bound nucleotides measured directly (●). At KCl concentrations above 1 M, the amount of bound ADP was about 1 mole per mole of myosin, but it decreased with decrease in the KCl concentration. The amount of bound ADP in 0.125 M KCl at 20 °C was 0.4 mole per mole of myosin (From Inoue and Tonomura[69], by courtesy of the Japanese Biochemical Society.)

The inserted figure shows the double reciprocal plot of the rate of the ATPase reaction in the steady state, v_0, against the concentration of ATP, [ATP]. The rate was measured from the liberation of $^{32}P_i$ using AT^{32}P as substrate in 0.5 M KCl at pH 7.8 and 0 °C. The concentration of myosin was 0.0024 mg ml^{-1}. The line represents $v_0(\text{min}^{-1}) = 0.11 + 0.33/\{1 + (1\,\mu M/[ATP])\}$. Four different preparations of myosin (○, △, $\times$, ●) were used. (From Inoue et al.[66], by courtesy of the Japanese Biochemical Society.)

where $^\circ$M is myosin of a changed conformation, which cannot react with ATP to form M_P^{ADP} but has no bound ADP or P_i. At low KCl concentrations the step $^\circ M \to M$ in the decomposition of M_P^{ADP} is rate-determining or its rate is comparable with that of the step $M_P^{ADP} \to {}^\circ M + ADP + P_i$. This mechanism was supported by the finding that when a sufficient amount of ATP was added to myosin in the presence of pyruvate kinase and PEP, 1 mole of ADP bound rapidly to 1 mole of myosin during the initial phase of the reaction, and then the amount of bound ADP decreased to the steady state level with an apparent rate constant equal to that of liberation of ADP and P_i from M_P^{ADP} (see the next section)[69].

Thus, the results on the myosin–ATPase reaction in the steady state can be easily explained by our mechanism[34], in which ATP is decomposed via the following two routes:

$$M + ATP \rightleftharpoons M_P^{ADP} \to {}^\circ M + ADP + P_i \to M + ADP + P_i \quad (4.1)$$

$$M(M_P^{ADP} \text{ or } {}^\circ M) + ATP \rightleftharpoons {}_{ATP}^{M} \to M + ADP + P_i \quad (4.2)$$

where $_{ATP}^{M}$ denotes the myosin–ATP complex, which is the complex for the main path of myosin–ATPase in the steady state at high ATP concentrations. ATP hydrolysis via route (4.1) is the main path of myosin–ATPase at low ATP concentrations with low K_m and V_{max} values, while ATP hydrolysis via route (4.2) is the main path at high ATP concentrations, with K_m and V_{max} values of 1 μM and 0.33 min^{-1}, respectively. We[70] reported previously that p-nitrothiophenol bound to myosin in the presence of ATP and high concentrations of Mg^{2+} ions, and that the ATPase activity in the steady state was affected only slightly, while the formation of M_P^{ADP} was completely inhibited by the p-nitrothiophenylation of myosin. Yagi et al.[71] prepared S-1N, which has a lower molecular weight than usual S-1, by digestion of S-1 with Nagarse, and showed that the formation of M_P^{ADP} was strongly inhibited in S-1N, while the ATPase activity of S-1N in the steady state was almost equal to that of usual S-1. These two results also suggest that ATP hydrolysis in the steady state at high ATP concentrations does not occur via M_P^{ADP}.

The findings that the amounts of $M_P^{ADP} + {}^\circ M$ and of $_{ATP}^{M}$ are both 1 mole per mole of myosin indicate heterogeneity of the two heads of the myosin molecule, although they do not indicate whether the active sites for formation of M_P^{ADP} and $_{ATP}^{M}$ are both located in the same head or in different heads.

4.3.2 Formation and decomposition of the reactive myosin–phosphate–ADP complex

As mentioned in Section 4.4, route (4.1) via the reactive myosin–phosphate–ADP complex, M_P^{ADP}, is the main path in the actomyosin–ATPase reaction, and the dissociation of actomyosin and recombination of myosin with actin are closely connected with the formation and decomposition of M_P^{ADP}. Thus, M_P^{ADP} has very important physiological functions in muscle contraction, and the step of its formation has been analysed by measuring the time courses of P_i liberation after stopping the reaction with TCA, liberation of

H^+ and change in the u.v. absorption spectrum of myosin during the initial phase of the reaction of myosin with ATP.

When the reaction of myosin with ATP is measured by quenching the reaction with TCA, there is a very rapid initial liberation of P_i called the 'initial burst'[72, 73], followed by a slower liberation in the steady state. This is because under usual experimental conditions the rate of formation of M_P^{ADP} is much greater than that of its decomposition, and P_i is liberated from M_P^{ADP} by quenching the reaction with TCA. The maximum amount of M_P^{ADP} was therefore measured from the size of the initial burst of P_i liberation, and found to be 1 mole per mole of myosin over a wide range of conditions[73, 74]. The size of the P_i initial burst was also measured[74] with myosin subfragments, and was about 1 and 0.5 mole per mole with HMM and S-1, respectively (Figure 4.3)*. These results suggest that M_P^{ADP} is formed in one

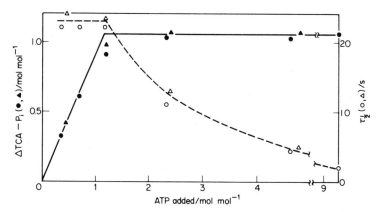

Figure 4.3 Dependences on the amount of ATP added of the amount of initial burst of P_i liberation and its rate for myosin and subfragment-1. 4.2 μM proteins, 2.8 M KCl, 10 mM $MgCl_2$, 20 mM Tris-HCl, pH 7.5, 0 °C. ●, amount of initial burst of P_i liberation for myosin (mol. wt. $= 4.8 \times 10^5$); ▲, amount of initial burst of P_i liberation for S-1 (for two moles of S-1 with mol. wt. $= 1.1 \times 10^5$). ○ and △ are the time for half of the maximum amount of initial burst of P_i liberation on addition of each concentration of ATP for myosin and S-1, respectively. The amount of initial burst of P_i liberation for both myosin and S-1 was equal to that of ATP added, when the latter was less than 1.1 mole per 2 moles of myosin heads. At higher ATP concentrations, the size of the initial burst of P_i liberation for both myosin and S-1 was independent of the amount of ATP added, and was 1.1 mole per 2 moles of myosin heads (From Hayashi and Tonomura[74], by courtesy of the Japanese Biochemical Society.)

* In many of our experiments the initial burst of P_i liberation amounted to 1.1, 1.1 and 0.55 moles per mole for myosin, HMM and S-1, respectively[34]. Values above the stoichiometric one may be due to unavoidable contamination with P_i liberated in the extra burst (*cf.* p. 139). The extra burst of initial liberation of P_i is observed especially at low Mg^{2+} concentrations, and amounts to several moles per mole. However, the results cannot exclude the possibility that our myosin preparation was a mixture of native myosin, showing 2 moles of initial burst per mole, and denatured myosin, showing no initial burst of P_i liberation, although this possibility is highly improbable.

of the two heads of myosin. On the other hand, Lymn and Taylor[65] later reported that the burst size of P_i liberation increased and approached 2 moles per mole of myosin with increase in the ATP concentration, but we[66] could not confirm this increase in the burst size.

We[75] measured the time course of liberation of H^+ in the myosin–ATPase reaction by measuring the change in optical density of cresol red using a stopped-flow method, and reported that 1 mole of H^+ per mole of myosin was rapidly liberated in the initial phase of the reaction. This result was later confirmed by other workers[76]. From the following three results, we concluded that there is also a very fast absorption of 1 mole of H^+ per mole of myosin, but that this occurs so rapidly that it is completed within the mixing period in the stopped-flow apparatus (cf. Figure 4.5). (i) At pH 8.0 the hydrolysis of 1 mole of ATP liberates[77] 1 mole of H^+, but liberation of the stoichiometric amount of H^+ was observed not only in the formation of M_P^{ADP} but also in its decomposition[78] into $M + ADP + P_i$. (ii) When the liberation of H^+ upon addition of a large amount of ATP to myosin was measured with a pH stat, the amount of H^+ liberated was found to increase linearly with time and there was not an initial fast period[75, 79]. (iii) When the solutions of ATP and the enzyme to be reacted were prepared at the same pH and the change in H^+ concentration was determined by the stopped-flow method, the pH value obtained by extrapolation of the generation of H^+ in the steady state to zero time was equal to the initial pH value of the solutions of the enzyme and ATP, but not to the pH when the rapid H^+ liberation started[75]. However, these observations are not direct evidence for the existence of an extremely rapid absorption of H^+, and the existence of H^+ absorption remains controversial.

Morita et al.[80, 81] and we[82] observed the change in u.v. absorption of HMM upon addition of ATP. Morita[81] also indicated that the change in the u.v. absorption spectrum induced by ATP occurs because particular tryptophan and tyrosine residues become buried in the non-polar region of the HMM molecule. We[83] measured the time courses of the initial burst of P_i liberation, the change in the u.v. absorption spectrum and the initial rapid liberation of H^+ under the same conditions. The liberation of H^+ and the change in the spectrum occurred concomitantly, as indicated in Figure 4.4. The rate of the initial burst of P_i liberation was lower than that of rapid H^+ liberation. Therefore we concluded that an intermediate (denoted as M_2ATP) which causes the rapid H^+ liberation and the change in the u.v. absorption spectrum is formed before M_P^{ADP}. Furthermore, we[84] measured the dependence of the rate of the rapid change in the spectrum on the ATP concentration. When the molar concentration of ATP was much higher than that of myosin, the time course of the change in the spectrum after adding ATP followed a first order kinetics, and the dependence of the pseudo-first order rate constant, k_f', on the ATP concentration was given by a Michaelis–Menten equation:

$$k_f' = k_f/\{1 + K_f/[ATP]\}$$

This result suggests that the intermediate (M_2ATP), of which the formation induces the change in the u.v. absorption spectrum, is formed via at least two steps. Denoting the intermediate formed before M_2ATP as M_1ATP, the following reaction mechanism was proposed[34, 84] for formation of M_P^{ADP}:

$$M + ATP \underset{k_{-1}}{\overset{k_1}{\rightleftharpoons}} M_1ATP \underset{k_{-2}}{\overset{k_2}{\rightleftharpoons}} M_2ATP \underset{k_{-3}}{\overset{k_3}{\rightleftharpoons}} M_P^{ADP}$$

(change in u.v. spectrum; (initial burst of
rapid H⁺ liberation) P_i liberation)

where k is the rate constant of each step and k_{-2} is much smaller than k_2. It was also concluded that there is a rapid equilibrium between $M + ATP$ and M_1ATP, since there was no lag phase in the time courses of the change in the u.v. absorption spectrum or the rapid H⁺ liberation in the initial phase*. The values of k_{-1}/k_1 and k_2 were found to be 0.2 mM and 30^{-1}, respectively, in 0.2 M KCl at 4 °C.

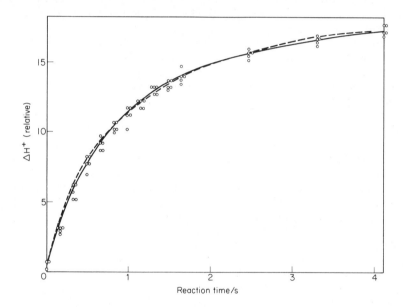

Figure 4.4 Initial rapid change in concentration of protons after adding ATP to myosin. 2 mg ml⁻¹ myosin, 5 μM ATP, 1.0 M KCl, 2 mM MgCl₂, 52 μM cresol red at pH 8.2, 21 °C. The solid line is drawn through the mean of 5 experimental measurements (○) of the initial liberation of H⁺, followed as the change in absorption of cresol red by a stopped-flow method. The broken line represents the mean of 5 measurements of the change in optical density of myosin at 293 nm. The liberation of H⁺ and change in the u.v. absorption spectrum correlated reasonably well, while the rate of initial burst of P_i liberation was slightly lower than those of rapid liberation of H⁺ and change in the u.v. absorption spectrum (From Tonomura *et al.*[83], by courtesy of the Japanese Biochemical Society.)

*The values of k_1 and k_{-1} can be estimated from the lag time, τ, and the Michaelis constant, K_f, for E_2ATP formation from the equations: $k_1 = 1/\tau\{[ATP] + K_f\}$; $k_{-1} = k_1 K_f - k_2$. By putting the values of K_f and τ into these equations, it was concluded that k_{-1} is much more than k_2.

The decomposition of M_P^{ADP} was measured after forming this complex by adding slightly less than the stoichiometric amount of ATP to myosin. The rate of ADP liberation from M_P^{ADP} was first measured by us[70] by adding the pyruvate kinase system at intervals after forming this complex, and by estimating the amount of free ADP from the time course of liberation of pyruvate (ADP + PEP $\rightarrow$ ATP + pyruvate). We[78] first assumed mistakenly that P_i was liberated from M_P^{ADP} after liberation of ADP. Recently, however, the rates of liberation of P_i and ADP from M_P^{ADP} were measured by Taylor *et al.*[85] using a gel-filtration method and by us[84] using a rapid-flow dialysis method, and it was shown that the rate of P_i liberation from M_P^{ADP} was almost equal to that of ADP liberation. The rate constants of liberation of ADP and P_i from M_P^{ADP}, k_d, were of the same order of magnitude as that of the V_{max} of ATPase in the steady state in the high concentration range of ATP*. For example, in 0.5 M KCl at pH 7.8 and 0°C the value of k_d was 0.75 min^{-1}, while the value of V_{max} was 0.44 min^{-1}, as described in Section 4.2.1. However, the dependences of k_d on KCl concentration and temperature differed considerably from those of the V_{max} of ATPase. Taylor *et al.*[85] thought that the liberation of ADP and P_i from M_P^{ADP} was accelerated by ATP itself, but this was disproved by our experiments using a rapid-flow dialysis method[84]. We also showed that the original u.v. absorption spectrum reappears when ADP and P_i are liberated from M_P^{ADP}.

As described in Section 4.3.1, M_P^{ADP} does not return to the initial enzyme form immediately after liberation of ADP and P_i from M_P^{ADP}. This was first concluded from the result that the rate constant of liberation of H^+ after forming M_P^{ADP} was several times less than those of ADP and P_i liberation from M_P^{ADP}, and was similar to the value of V_{max} of ATPase in the steady state in the low concentration range of ATP (Figure 4.5)[78, 83]. This conclusion was also supported by measurement of the rate of recovery of the ability to show the initial burst of P_i liberation after formation of M_P^{ADP} by the reaction of ATP with myosin in a 1:1 molar ratio[86]:

$$M_P^{ADP} \rightarrow {}^\circ M + ADP + P_i \rightarrow M + ADP + P_i + H^+$$

(recovery of u.v.   (slow liberation of H^+;
spectrum)   recovery of initial activity
                    for P_i burst)

where $^\circ M$ is myosin of a changed conformation, which cannot show the initial burst of P_i liberation but has no bound ADP or P_i. As already mentioned in Section 4.3.1, in the steady state of myosin–ATPase, less than the stoichiometric amount of ADP was bound to myosin. Since the rate constant of M_P^{ADP} formation is extremely high in comparison with that of decomposition of M_P^{ADP}, this means that in route (4.1) myosin mainly takes the form of M_P^{ADP} or $^\circ M$†.

---

* This point is discussed on p. 137.

†In this mechanism the amount of M_P^{ADP} was calculated from the ratio of the rate constants of step $M_P^{ADP} \rightarrow {}^\circ M$ and step $^\circ M \rightarrow M$ as 0.2–0.3 mole per mole of myosin in 0.5 M KCl at 0°C, while the amount of ADP-binding was observed to be about 0.7 mole per mole. The difference between the observed and calculated values might be attributed to ADP-binding other than M_P^{ADP}.

From kinetic results on the myosin–ATPase reaction in the steady state and on the formation and decomposition of M_P^{ADP} described in Section 4.3.1 and 4.3.2, respectively, the following mechanism is proposed for the myosin–ATPase reaction:

$$M + ATP \rightleftharpoons M_1ATP \rightleftharpoons M_2ATP \rightleftharpoons M_P^{ADP} \rightarrow {}^{\circ}M + ADP + P_i$$
$$\rightarrow M + ADP + P_i \tag{4.1}$$

$$M(M_P^{ADP} \text{ or } {}^{\circ}M) + ATP \rightleftharpoons {}^{M}_{ATP} \rightarrow M + ADP + P_i \tag{4.2}$$

In 0.5 M KCl at $0\,^{\circ}$C the values of V_{max} and K_m for ATP hydrolysis via route (4.1) were calculated to be 0.11 min^{-1} and less than 0.1 μM, respectively, from the rate constant of each step, while those via route (4.2) were calculated as 0.33 min^{-1} and 1 μM, respectively.

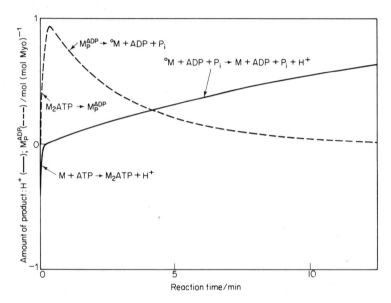

Figure 4.5 Schematic view of the courses of change in the amount of the reactive myosin–phosphate–ADP complex and that of H$^+$, when the reaction was started by adding ATP to myosin in a 1 : 1 molar ratio. 0.5 M KCl, 2 mM MgCl$_2$, pH 7.8, $0\,^{\circ}$C. – – –, amount of the reactive myosin–phosphate–ADP complex, M_P^{ADP}; ———, absorption and liberation of H$^+$. When ATP is added to myosin in a molar ratio of 1 : 1, almost all of it is hydrolysed via route (4.1), $M + ATP \rightleftharpoons M_1ATP \rightleftharpoons M_2ATP \rightleftharpoons M_P^{ADP} \rightarrow {}^{\circ}M + ADP + P_i \rightarrow M + ADP + P_i$. Various elementary steps of the myosin–ATPase reaction can be observed separately by measuring the changes in M_P^{ADP} and H$^+$, as indicated in the figure.

However, the results on the transient phase of the ATPase reaction can also be explained by supposing[34] that ATP only decomposes via route (4.1) and that ATP accelerates the conversion of $^{\circ}M$ to M at ATP concentrations above 0.5 μM:

low ATP conc.:

$$M + ATP \rightleftharpoons M_P^{ADP} \rightarrow {}^{\circ}M + ADP + P_i \nrightarrow M + ADP + P_i$$

high ATP conc.:

$$\text{accelerated by ATP } (\gtrsim 0.5\,\mu M)$$

$$M + ATP \rightleftharpoons M_P^{ADP} \nrightarrow {}^{\circ}M + ADP + P_i \longrightarrow M + ADP + P_i$$

The finding that the rates of liberation of ADP and P_i from M_P^{ADP} are of the same order of magnitude as the value of V_{max} of the ATPase reaction in the steady state in a high concentration range of ATP can easily be explained by this mechanism, since in this scheme the rate-determining step of the ATPase reaction at low concentrations of ATP is the conversion of ${}^{\circ}M$ to M, while at high concentrations of ATP the rate-determining step is the step from M_P^{ADP} to ${}^{\circ}M + ADP + P_i$. However, this mechanism is not supported by results on the amount of ADP bound to myosin during the ATPase reaction. It is expected from this mechanism that the amount of M_P^{ADP} in the steady state at low ATP concentrations is less than stoichio-metric $\left(\sum [M] \approx [M_P^{ADP}] + [{}^{\circ}M]\right)$, while at high ATP concentrations it is 1 mole per mole of myosin $\left(\sum [M] \approx [M_P^{ADP}]\right)$, since the amounts of M, M_1ATP and M_2ATP are expected to be negligible if the amount of ATP added is more than that of the enzyme. However, as mentioned in Section 4.3.1, the amount of bound ADP was constant and unaffected by the amount of ATP added, when more than the stoichiometric amount of ATP was added. The amount of bound ADP depended on the KCl concentration, and was 1 mole per mole of myosin at KCl concentrations above 1 M, while at or below 0.5 M KCl it was lower than the stoichiometric amount. This mechanism is also not supported by the findings mentioned in Section 4.3.1 that p-nitrothiophenyl myosin and Nagarse-treated S-1 showed no initial burst of P_i liberation but had full ATPase activity in the steady state at high ATP concentrations.

The mechanism of myosin–ATPase proposed above was deduced mainly from results obtained in the presence of concentrations of 0.5 M KCl or less and of more than 1 mM Mg^{2+}. The mechanism has to be modified when the experimental conditions are changed. For example, at concentrations of above 1.5 M KCl the rate of liberation of H^+ after forming M_P^{ADP} was almost equal to both the rate of ADP liberation from M_P^{ADP} and the rate of the ATPase reaction in the steady state at high ATP concentrations[83]. These results indicate that at very high KCl concentrations ATP hydrolysis via route (4.1) is the main path and the rate-determining step of route (4.1) is the step from M_P^{ADP} to ${}^{\circ}M + ADP + P_i$. This conclusion is also consistent with the finding that the amount of ADP bound to myosin during the ATPase reaction was 1 mole per mole of myosin at high KCl concentrations (Section 4.3.1)[69].

Taylor et al.[85] and Malik and Martonosi[87, 88] claimed that the kinetic properties of the simple myosin–ADP complex, M·ADP, which is formed by adding ADP to myosin, are the same as those of M_P^{ADP}. However, it is well known that the u.v. absorption spectrum[81, 82] and the fluorescent spectrum[89, 90] of myosin and the e.p.r. spectrum[91-93] of spin-labelled myosin in the presence of ATP are all different from those of M·ADP. Trentham et al.[94] utilised the change in the u.v. absorption spectrum due to the binding of HMM with SH analogues of ATP (SH-ATP, SH-ADP) to measure the rate of product liberation, and showed that the rate constant of liberation of SH-ADP from HMM·SH-ADP is much higher than that of SH-ADP from HMM_P^{SH-ADP}. Using a rapid-flow dialysis method we[84] showed that the rate constant of ADP liberation from M·ADP is much higher than that of ADP liberation from M_P^{ADP}. Furthermore, the initial burst of P_i liberation of the stoichiometric amount was observed, even when ATP was added to the M·ADP complex formed by adding a sufficient amount of ADP to myosin, indicating that the myosin–ADP complex is not the most stable intermediate in route (4.1) of the myosin–ATPase reaction[84]. Recently, the rate constant, k_d, of dissociation of ADP from the HMM–ADP complex was measured by the following four different methods[95]. (i) The apparent rate constant of formation of the HMM–ADP complex after adding ADP to HMM, k, which was equal to $k_d + k_b[ADP]$, was measured by the change in u.v. absorption of HMM. Then the value of k_d was obtained as the value of k extrapolated to $[ADP] = 0$. (ii) The value of k_d was calculated as the product of the dissociation constant, K_D, and the binding rate constant, k_b, obtained from k. (iii) The value of k_d was determined from the rate of change in the u.v. absorption after adding a sufficient amount of PP_i to the HMM–ADP complex. (iv) It was also determined using ATP, instead of PP_i, as a displacing agent. All the values of k_d obtained were more than 10 times that of liberation of ADP from M_P^{ADP} at all temperatures used.

4.3.3 Structure of the reactive myosin–phosphate–ADP complex

It is now well established that the reactive myosin–phosphate–ADP complex, M_P^{ADP}, is the key intermediate in muscle contraction, and elucidation of the biochemical structure of this complex is essential for clarification of the molecular mechanism of energy transduction in muscle.

It is important to know the structural change of myosin upon reaction with ATP. The changes in the u.v. absorption spectrum and the fluorescent spectrum of myosin on addition of ATP were concluded to be induced by local conformational changes around the active site induced by ATP. It was also concluded by Werber et al.[89] and by Morita[90] that upon reaction of myosin with ATP at least two tryptophan residues are buried in the non-polar region of the myosin molecule. We[96] have recently studied the kinetic properties of P_i liberation and change in the u.v. and fluorescent spectra of HMM during its reaction with a fluorescent analogue of ATP, 1-N^6-etheno-ATP (εATP). Our studies showed that $HMM_P^{εADP}$ is formed by the reaction of HMM with εATP, and that during its formation at least two tryptophan residues are buried in the non-polar region of the HMM molecule. We discovered that energy transfer from tryptophan residues to εADP bound to HMM occurs efficiently, and the transfer occurs only from tryptophan residues accessible to KI, not from the residues buried in the non-polar region. This suggests that the tryptophan residues which are buried on adding εATP are located far from the bound εADP, since it is rather improbable that the planes of vibration of two tryptophan residues which are buried on adding εATP are both oriented perpendicularly to that of bound εADP. In other words, the structural change of myosin upon reaction with ATP extends over a rather wide region of the myosin molecule. However, it should be noted that the change in u.v. absorption of myosin is caused by formation of M_2ATP and is conserved in M_P^{ADP}, as mentioned in the previous section.

It is controversial whether phosphate binds to myosin in M_P^{ADP} through a covalent or non-covalent bond. At first we thought that, in M_P^{ADP}, phosphate is bound to myosin as a TCA-unstable acyl phosphate for the following reasons[34]. (i) No H^+ was liberated in the overall reaction[75] during the formation of M_P^{ADP}. (ii) p-Nitrothiophenol, which is known as a nucleophilic reagent, binds to a specific glutamic acid residue of myosin only in the presence of ATP and high concentrations of Mg^{2+} ions, and this p-nitrothiophenylation of myosin suppresses the initial burst of P_i liberation[70], without altering the ATPase activity in the steady state*.

Later Sartorelli et al.[98] reported that ^{18}O was not incorporated into P_i upon decomposition of M_P^{ADP} in $H_2^{18}O$ at neutral pH, suggesting that the bond between P_i and myosin in this complex is non-covalent. On the other hand, we[86] found that a P-exchange reaction occurs between the reaction intermediate of myosin–ATPase in the initial phase and γ-^{32}P-labelled ATP. However, only 10–20% of M_P^{ADP} undergoes the exchange reaction, even in 2.8 M KCl, and with the usual KCl concentration ($\leqslant 0.6$ M) scarcely any M_P^{ADP} undergoes the exchange reaction.

In connection with the structure of M_P^{ADP}, it is interesting to note that the size of the initial burst of P_i liberation reaches over ten moles per mole of myosin when the concentration of Mg^{2+} ions is reduced[73] to a few μM. This phenomenon was called the 'extra-burst' of P_i liberation. Our analysis of the extra-burst showed that, at low concentrations of Mg^{2+} ions, the direct decomposition of a reaction intermediate, which is formed transiently before formation of M_P^{ADP}, occurs in the initial phase of the reaction until the intermediate is finally converted into M_P^{ADP}. We[83] also found that p-nitrothiophenylation of myosin in the presence of ATP and a low concentration of Mg^{2+} ions completely eliminates the extra-burst of P_i liberation, but does not affect the steady state rate of ATPase or the initial stoichiometric burst in the presence of high concentrations of Mg^{2+}. From these results we concluded that phosphoryl myosin, in which phosphate binds to myosin as a TCA-unstable acyl phosphate, $M_{\sim P}^{ADP}$, is formed transiently before formation of the myosin–phosphate–ADP complex, $M_{\cdot P}^{ADP}$, and at low concentrations of Mg^{2+} ions direct decomposition of $M_{\sim P}^{ADP}$ occurs rapidly until all of $M_{\sim P}^{ADP}$ is converted into $M_{\cdot P}^{ADP}$, while at high concentrations of Mg^{2+} ions the rate of step $M_{\sim P}^{ADP} \rightarrow M_{\cdot P}^{ADP}$ is much larger[34, 70] than that of direct decomposition of $M_{\sim P}^{ADP}$. Based on this conclusion, the reactive myosin–phosphate–ADP complex, M_P^{ADP}, contains both $M_{\sim P}^{ADP}$ and $M_{\cdot P}^{ADP}$.

*Wolcott and Boyer[97] recently reported that p-nitrothiophenol binds to SH groups of myosin. However, in their experiments p-nitrothiophenylation of myosin was not specifically dependent on the presence of Mg^{2+}–ATP, contrary to our findings, and they did not clarify the effect of the modification on the initial burst of P_i liberation. Therefore, it is possible that their p-nitrothiophenyl myosin was produced as a result of a side reaction.

140

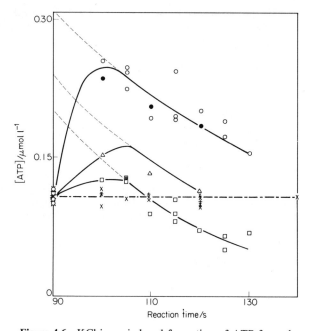

Figure 4.6 KCl-jump induced formation of ATP from the
reactive myosin–phosphate–ADP complex. The reaction was
started by adding 3 μM AT^{32}P to a solution containing 3
mg ml^{-1} HMM, 1.5 M KCl, 2 mM MgCl$_2$ and 50 mM Tris-
HCl (pH 7.8) at 10 °C. After 90 s, when decrease in the
amount of AT^{32}P had almost stopped, 0.5 ml of the reaction
mixture was rapidly poured into 14.5 ml of solution containing 2 mM MgCl$_2$ and 50 mM Tris-HCl (pH 7.8) at 0 (O),
10 (△) and 20 °C (□) to decrease the KCl concentration
rapidly from 1.5 to 0.05 M, and at intervals the amount of
AT^{32}P was measured. When 0.5 ml of the reaction mixture
was poured into 14.5 ml of a solution containing 1.5 M KCl,
the amount of AT^{32}P did not change either at 0 (×) or at
20 °C (+). ● shows the time course of change in the amount
of ATP when the final solution contained 1 mM of unlabelled ATP and 0.05 M KCl at 0 °C. The amount of AT^{32}P
measured after the KCl-jump was expressed as μM in the
original volume before the KCl-jump by multiplying the
observed concentration in the diluted solution by 30. The
solid line for decay of the amount of ATP formed by the
KCl-jump was the best fit, when the rate constant of liberation of ADP and P$_i$ from M$_P^{ADP}$ (1, 1.5 and 2 min^{-1} at 0, 10
and 20 °C) was used as the rate constant of decay in AT^{32}P.
The amount of ATP formed by the KCl-jump was estimated
by extrapolating the solid line to the time of the KCl-jump.
The equilibrium constants of the step M$_2$ATP ⇌ M$_P^{ADP}$, calculated from the amounts of ATP formed, were 6.1 and 10.0 in
0.05 M KCl at 0 and 20 °C, respectively. They were in good
agreement with the values estimated by the kinetic method
of Bagshaw and Trentham, i.e. 5.8 and 9.0 at 0 and 20 °C
(From Inoue *et al.*[100], by courtesy of the Japanese Biochemical
Society.)

Analysis of the equilibrium between M_P^{ADP} and M_2ATP gave very interesting information on the structure of M_P^{ADP}. The existence of an equilibrium in this step was suggested by Bagshaw and Trentham[99] from the findings that when excess S-1 was mixed with $AT^{32}P$ in 0.05 M KCl, about 10% of the added $AT^{32}P$ was present at the time of completion of the initial burst of P_i liberation*, and that the apparent rate constant of decay of the $AT^{32}P$ remaining was equal to that of decomposition of M_P^{ADP} into $^\circ M + ADP + P_i$, and was unaffected by subsequent addition of unlabelled ATP. We[100] confirmed their results, and showed the equilibrium more directly, using the dependence of the equilibrium constant of this step on KCl concentration. When M_P^{ADP} was formed by mixing excess HMM with ATP in 1.5 M KCl and then the KCl concentration was rapidly decreased from 1.5 to 0.05 M (KCl jump), ATP was formed from M_P^{ADP} (Figure 4.6). The amount of ATP formed was almost equal to that of M_2ATP, calculated from the change in the equilibrium constant estimated by the kinetic method. The apparent rate constant of decay of the ATP thus formed was also equal† to that of the step from M_P^{ADP} to $^\circ M + ADP + P_i$.

The thermodynamic properties of the equilibrium between M_2ATP and M_P^{ADP} were studied using the KCl-jump and the kinetic method. The value[100] of ΔS° was 16–19 cal K^{-1} mol^{-1}, and was scarcely affected by the KCl concentration, while ΔH° decreased from 5.1–4.4 to 3.0 kcal mol^{-1} with increase in the KCl concentration from 0.05–0.1 to 0.5 M. It is uncertain whether the increase in entropy in the step $M_2ATP \rightleftharpoons M_P^{ADP}$ is due to conformational changes in myosin, release of bound water or other factors, but the positive value of ΔS° excludes the possibility that ΔG for hydrolysis of ATP in the step $M_2ATP \rightleftharpoons M_P^{ADP}$ is stored in the conformational entropy of myosin. The positive value of ΔH° suggests that the energetic and structural states of bound phosphate in M_P^{ADP} are very different from those of P_i in solution. This is also supported by the finding that the step $M_1ATP \rightleftharpoons M_2ATP$, but not the step $M_2ATP \rightleftharpoons M_P^{ADP}$, is accompanied by H^+ liberation (cf. p. 134). Recently, we[101] measured the thermodynamic functions of the step between the second enzyme–ATP complex and phosphorylated intermediate, $E_2ATP \rightleftharpoons E_{\sim P}^{ADP}$, in the Na^+,K^+-dependent ATPase reaction. The values of ΔH° and ΔS° in the presence of a sufficient amount of NaCl and the absence of KCl were 4.3 kcal mol^{-1} and 15.6 cal K^{-1} mol^{-1}, respectively, which are very similar to those observed in the myosin–ATPase reaction. These results strongly suggest the similarity of the structure of M_P^{ADP} in the myosin–ATPase reaction and $E_{\sim P}^{ADP}$ in the Na^+,K^+-dependent ATPase reaction before the addition of TCA, although in the myosin–ATPase reaction the phosphate of M_P^{ADP} is liberated as P_i on adding TCA (Section 4.3.2), while in the Na^+,K^+-dependent ATPase reaction the phosphate remains bound to the enzyme as an acyl phosphate after TCA denaturation[102, 103].

---

*Since the Michaelis constant of route (4.1) was much less than that of route (4.2), ATP hydrolysis via route (4.2) could be neglected under the conditions where the molar concentration of the ATPase active site was much higher than that of ATP added (cf. Figure 4.2).

† The equilibrium between M_2ATP and M_P^{ADP} shifts greatly to the M_P^{ADP}-side, and the maximum amount of M_2ATP is about 10% of that of M_P^{ADP}. Therefore the equilibrium could be neglected in discussion on the amount of M_P^{ADP} and the rate of its decomposition described in the previous section.

4.4 REACTION MECHANISM OF ACTOMYOSIN–ATPase

4.4.1 Dissociation of actomyosin by ATP

As described in Section 4.1, the ATPase activity of myosin is markedly activated by actin[2], and actomyosin–ATPase is directly coupled with muscle contraction. There was some uncertainty on whether 1 or 2 moles of actin monomer bound with 1 mole of myosin[34]. However, our recent studies[104] using light-scattering, ultracentrifugation and electron microscopy established that 1 mole of myosin binds with 2 moles of actin monomer. As mentioned in Section 4.2, each of the two heads of myosin binds with actin monomer[104]. X-Ray diffraction and electron microscopy also showed that in living muscle the head parts of the myosin molecule bind to F-actin and constitute a cross-bridge[34]. Moreover, the molar ratio of myosin to actin monomer in myofibrils was found[105] to be about 1:7, which is much less than the binding ratio in the isolated system. This small ratio is considered to be the structural basis for the independent movement of many cross-bridges on one thick filament in muscle contraction[106].

At high ionic strength, ATP induces the dissociation of actomyosin into F-actin and myosin[2]. The molar concentration of ATP necessary to induce this dissociation was the same as that of the myosin moiety in actomyosin, and did not depend on the amount of F-actin, indicating that ATP reacts with the myosin moiety[107]. The dissociation of actomyosin was also induced by PP_i. The mechanism of dissociation of actomyosin induced by PP_i is more readily understandable than that by ATP, since PP_i is not decomposed by actomyosin[59, 108]. As mentioned in Section 4.2, one mole of PP_i binds to 1 mole of myosin moiety in actomyosin. We[108] compared the amount of binding of PP_i to myosin with the extent of dissociation of actomyosin under the same conditions, and also measured the rates of dissociation after adding PP_i and recombination of actomyosin after rapid hydrolysis of PP_i with pyrophosphatase. The following mechanism of reaction of actin, myosin and PP_i was deduced from our kinetic and thermodynamic analyses[108]:

$$AM + PP_i \rightleftharpoons AM{\cdot}PP_i \rightleftharpoons A + M{\cdot}PP_i$$

In this scheme, A and M denotes the reaction units of actin and myosin, while $AM{\cdot}PP_i$ and $M{\cdot}PP_i$ denote the complexes of actomyosin and myosin with PP_i. The most important point in this mechanism is that there is a complex of actomyosin with PP_i, which is in equilibrium with $A + M{\cdot}PP_i$. On the other hand, the following mechanism was tacitly assumed by many workers for the dissociation:

$$AM \rightleftharpoons A + M$$
$$M + PP_i \text{ (or ATP)} \rightleftharpoons M{\cdot}PP_i \text{ (M}{\cdot}\text{ATP)}$$

In this mechanism the equilibrium of the first step is shifted to the right by the binding of PP_i or ATP to myosin but not to actomyosin. However, this mechanism can be excluded by the fact that the rate constant of dissociation of actomyosin into F-actin and myosin, calculated as the product of the dissociation constant and the rate constant of binding of myosin with F-actin, is too low to explain the rapid dissociation of actomyosin induced by PP_i or ATP.

We[107] showed that the reactive myosin–phosphate–ADP complex, M_P^{ADP}, is formed upon the reaction of actomyosin with ATP, and that the amount of myosin dissociated from actomyosin is equal to that of M_P^{ADP} formed by the reaction. We compared the rate of dissociation of actomyosin upon addition of low concentrations of ATP to actomyosin with that of formation of M_P^{ADP}. At high concentrations of Mg^{2+} ions the two rates were almost equal[107], but at low concentrations of Mg^{2+} ions the rate of M_P^{ADP} formation was slightly higher than that of dissociation of actomyosin[109]. Based on these results, we proposed the following mechanism for the dissociation of actomyosin induced by ATP:

$$AM + ATP \underset{(1)}{\rightleftharpoons} AM_1ATP \underset{(2)}{\rightleftharpoons} AM_2ATP \underset{(3)}{\rightleftharpoons} AM_P^{ADP} \underset{(4)}{\rightleftharpoons} A + M_P^{ADP}$$

We[107] also showed that at high ionic strength the recombination of F-actin with M_P^{ADP} occurs after transition of M_P^{ADP} into $^\circ M + ADP + P_i$, and that the rate of combination of $^\circ M$ with F-actin is much higher than that of conversion of $^\circ M$ into M. It is interesting to note that the amounts of ATP necessary to cause the maximum dissociation of actomyosin, acto-HMM and acto-S-1 were 1, 2 and 2 moles per 2 moles of the head of myosin, i.e. per mole of myosin and HMM, and 2 moles of S-1, respectively[104]. This result indicates that there is a strong interaction between the two heads of myosin, at least when myosin is bound with actin, and that this interaction is modified by tryptic digestion of myosin to form HMM or S-1.

Since the amounts of ATP and PP_i necessary to cause the dissociation of actomyosin are both 1 mole per mole of myosin moiety in actomyosin, we may consider that both ATP and PP_i bind to the same site in the myosin molecule. Then it is reasonable to assume that ATP can dissociate actomyosin not only by formation of M_P^{ADP} but also by simple binding to the myosin moiety, as is the case of PP_i:

$$AM + ATP \underset{(1)}{\rightleftharpoons} AM_1ATP \quad \overset{(5)}{\nearrow} \quad A + M_1ATP \overset{(6)}{\rightleftharpoons} A + M_2ATP \quad \overset{(7)}{\searrow} \quad A + M_P^{ADP}$$
$$\underset{(2)}{\searrow} AM_2ATP \underset{(3)}{\rightleftharpoons} AM_P^{ADP} \overset{(4)}{\nearrow}$$

where the dissociation of actomyosin is induced by low concentrations of ATP via steps (2), (3) and (4), while the dissociation by PP_i occurs via steps (5) and (6)*. It is well known that the rate of dissociation of actomyosin increases with increase in the ATP concentration[34]. This result can be explained by supposing that the dissociation of actomyosin by ATP is induced by formation of $M_{\sim P}^{ADP}$ (cf. Section 4.3.3) and that ATP at high concentrations accelerates the reaction from $M_{\sim P}^{ADP}$ to $M_{\cdot P}^{ADP}$, as we proposed previously [34, 35],

*If we assume only the route via $A + M_1ATP$ for the dissociation, we cannot explain the dependence of the actomyosin–ATPase activity on the extent of the dissociation, which will be mentioned in the next section.

or by supposing that, at high concentrations of ATP, the dissociation via steps (5), (6) and (7) becomes the main path. Lymn and Taylor[110] reported that, at extremely high concentrations of ATP, the rate of dissociation of actomyosin by ATP was much higher than that of formation of M_P^{ADP}. It is doubtful whether the high rate constant for the dissociation of actomyosin reported can be measured quantitatively by a stopped-flow method, since actomyosin has high viscosity and is easily fragmented by shearing stress. However, in the above mechanism we have adopted the latter of the two mechanisms, accepting Lymn and Taylor's conclusion qualitatively*.

4.4.2   Two-route mechanism of actomyosin–ATPase

We[34] concluded that the reactive myosin–phosphate–ADP complex, M_P^{ADP}, is the reaction intermediate in the actomyosin–ATPase reaction, and that at low ionic strength F-actin accelerates the decomposition of M_P^{ADP}, without affecting its formation rate. The reasons for this conclusion were as follows. (i) As described in Section 4.3.2, the complete decomposition of M_P^{ADP} into $M + ADP + P_i$ is accompanied by the liberation of 1 mole of H^+ per mole of myosin, and this H^+ liberation is greatly accelerated by adding F-actin at low ionic strength[78]. (ii) Modification of myosin with p-nitro-thiophenol, under conditions where the initial stoichiometric burst occurred, did not affect ATPase activity in the steady state, but completely inhibited the initial burst of P_i liberation. The ATPase activity of the p-nitrothiophenyl myosin thus formed was not accelerated by F-actin[109, 111]. Both the initial burst of P_i liberation and the activation of ATPase by F-actin were coincidently suppressed by digestion of S-1 with Nagarse[71] and modification of the SH_1 groups of myosin with NEM[112]. (iii) When F-actin was added to the myosin–ATP system afterwards, the activation of ATPase by F-actin occurred immediately without any lag time[113]. (iv) It is well established[112, 114] that treatment of myosin with CMB and then with β-mercaptoethanol does not essentially alter the myosin–ATPase activity either in the steady state or in the initial phase, but that actomyosin reconstituted from CMB-treated myosin is resistant to dissociation on adding ATP. Therefore we compared the rate of formation of M_P^{ADP} of myosin with the rate of actomyosin–ATPase, using myosin treated with CMB and β-mercaptoethanol. As shown in Figure 4.7, at low ionic strength the two rates were almost equal over a wide range of ATP concentrations[66]. This result showed conclusively that M_P^{ADP} is the reaction intermediate in the actomyosin–ATPase reaction and F-actin greatly accelerates the step of regeneration of myosin from M_P^{ADP}, without affecting the rate of its formation.

Therefore the next problem is the mechanism of acceleration of decomposition of M_P^{ADP} by F-actin. Straub and Feuer[115] discovered that 1 mole of ATP is bound to 1 mole of G-actin, and that on transformation of G-actin to F-actin the ATP is dephosphorylated and F-actin contains bound ADP. Based on these findings several workers have assumed that nucleotide

*According to this mechanism the rate of step (5) is accelerated by a high concentration of ATP.

bound to actin has some important function in muscle contraction. Mom-maerts[116] proposed that ATP hydrolysis accompanied by $G \rightleftharpoons F$ trans-formation of actin induces muscle contraction. This type of mechanism seems to be supported by the findings of Asakura et al.[117] that actin catalyses a very slow hydrolysis of ATP on ultrasonication and of Szent-Györgyi[118] that ADP bound to F-actin exchanges with external nucleotide during super-precipitation of actomyosin with ATP. However, the so-called ATPase reaction of actin on ultrasonication was later shown to be due to the fragmen-tation of F-actin by sonication, followed by the recombination of fragments accompanied by ATP hydrolysis[119]. Furthermore, it was shown that not only ATP but also ADP, which does not induce superprecipitation of actomyosin, exchanges with ADP bound with F-actin, and that the exchange reaction of nucleotide occurs as a result of the binding of F-actin with the myosin filament[120]. Using ADP-free F-actin Bárány et al.[121] and we[122] showed that CTP and deoxy-ATP were not incorporated into F-actin, while their hydroly-sis by myosin was markedly activated by F-actin, and that they induced typical superprecipitation of actomyosin. Therefore, it is now generally accepted that nucleotide bound to actin is not required for the actomyosin–ATPase reaction and muscle contraction.

The ATPase activity of purified actomyosin is inhibited by high concen-trations of ATP (substrate inhibition), and decreases with increase in the ATP concentration to a level which is higher than that of myosin-ATPase[34].

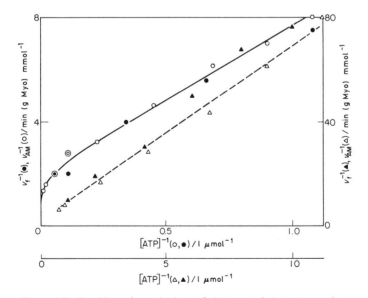

Figure 4.7 Double reciprocal plots of the rate of the actomyosin–ATPase reaction and the rate of the initial burst of P_i liberation from myosin–ATPase against ATP concentration. 50 mM KCl, 2 mM $MgCl_2$, 20 mM Tris-maleate, pH 7.0, 25 °C. Myosin was used after treatment with CMB and β-mercaptoethanol to strengthen its binding to F-actin. $\bigcirc$, $\triangle$, rate of actomyosin–ATPase, v_{AM}; $\bullet$, $\blacktriangle$, rate of initial burst of P_i liberation from the myosin–ATP system, v_f (From Inoue et al.[66], by courtesy of the Japanese Biochemical Society.)

In the presence of the regulatory protein (a tropomyosin and troponin complex), the ATPase activity of actomyosin at high concentrations of ATP is dependent on the Ca^{2+} concentration[32]. In the absence of Ca^{2+} ions the activation of myosin–ATPase by actin is completely inhibited; this state is called the relaxed state.

At high concentrations of ATP the decomposition of M_P^{ADP} is rate-determining in the ATPase reaction, since the rate of M_P^{ADP} formation is extremely high. Therefore we[123] compared the rate of the acto-H-meromyosin–ATPase reaction in the steady state at high concentrations of ATP, v_0, with that of binding of F-actin with HMM_P^{ADP}, which was formed by adding ATP to HMM in a 1:1 or 2:1 molar ratio. The maximum rate of acto-H-meromyosin–ATPase was obtained by extrapolating the double reciprocal plot of v_0 against F-actin concentration to an infinite concentration of F-actin. The extent of dissociation of acto-H-meromyosin in the presence of high concentrations of ATP and the rate of the H-meromyosin–ATPase reaction were also measured under the same conditions. In the presence of various concentrations of F-actin, the value of v_0 was obtained by the following equation from the rate of the H-meromyosin–ATPase reaction, k_m, in the steady state, the maximum rate of acto-H-meromyosin–ATPase, k_8, the rate of binding of F-actin with HMM_P^{ADP}, k_9, and the extent of dissociation of acto-H-meromyosin in the presence of high concentrations of ATP, a (Figure 4.8):

$$v_0 = k_m + (1 - a)k_8 + ak_9$$

The value of k_8 was about $4\,s^{-1}$ in 50 mM KCl at 23 °C.

This result indicates that in the actomyosin–ATPase reaction ATP is hydrolysed via the route in which the binding of F-actin with HMM_P^{ADP} is rate-determining and the route in which ATP is hydrolysed by undissociated actomyosin:

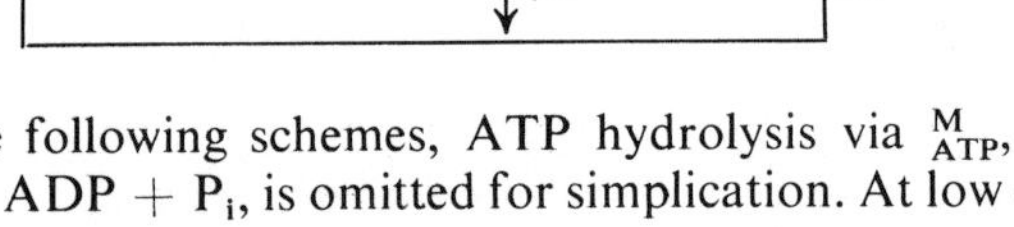

$$
\begin{array}{ccc}
P_i & & \\
+ & & \\
ADP & (1-a) & (a) \\
+ & & \\
AM + ATP \rightleftharpoons AM_P^{ADP} & \rightleftharpoons & A + M_P^{ADP} \\
\end{array}
$$

（8）　　　　（9）

In this and the following schemes, ATP hydrolysis via $\overset{M}{_{ATP}}$, $M + ATP \rightleftharpoons \overset{M}{_{ATP}} \overset{km}{\longrightarrow} M + ADP + P_i$, is omitted for simplication. At low concentrations of F-actin, actomyosin dissociated almost completely ($a \approx 1$), and the rate of actomyosin ATPase, v_0, was given as $k_m + k_9$. This is consistent with the previous report[124] that the ATPase activity of HMM was accelerated by F-actin, in spite of the complete dissociation of acto-H-meromyosin. On the other hand, at high concentrations of F-actin the rate of acto-H-meromyosin–ATPase, v_0, in the steady state was much higher than that of binding of HMM_P^{ADP} with F-actin ($v_0 > k_m + k_9$). These results provide direct evidence for our two-route mechanism of actomyosin–ATPase[34], and are inconsistent with the mechanism proposed by Lymn and Taylor[110] in which

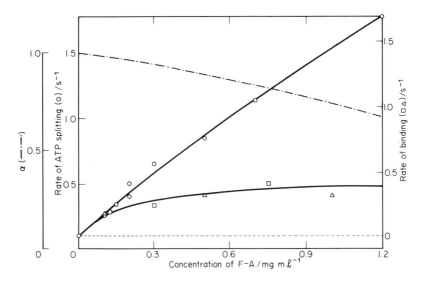

Figure 4.8 Dependences on F-actin concentration of the rate of acto-H-mero-myosin, the rate of binding of F-actin with the reactive H-meromyosin–phos-phate–ADP complex and the extent of dissociation of acto-H-meromyosin in the steady state. 50 mM KCl, 2 mM MgCl$_2$, 10 mM Tris-HCl, pH 7.8, 20 °C. The rate of acto-H-meromyosin–ATPase activity in the steady state (○) was measured in 0.136 mg ml^{-1} HMM and 0.33 mM ATP. The double reciprocal plot of the rate, v_0, versus F-actin concentration gave a straight line. The value of k_8 was obtained as v_0 at an infinite F-actin concentration. ———, level of ATPase of HMM alone, k_m. The rate of binding of F-actin with HMM$_P^{ADP}$ was measured from the increase in light-scattering intensity using a stopped-flow method. F-Actin was added at intervals after adding ATP to 0.136 mg ml^{-1} HMM in a 1:1 (△) or 2:1 (□) molar ratio. The increment of light-scattering intensity was expressed as the sum of the rapid and slow increments of light-scattering intensity, and the percentage of rapid increment increased gradually with the interval between the time of mixing ATP with HMM and that of adding F-actin. The rate constant of this increase (0.048 s^{-1}) was the same as that of liberation of ADP and P$_i$ from HMM$_P^{ADP}$. Therefore, the slow increase in light-scattering intensity was concluded to be due to the binding of F-actin with HMM$_P^{ADP}$, and the apparent rate constant, k_9, of the binding was obtained by plotting the logarithm of the increment in light-scattering intensity against time. Values of a (—·—·—) were calculated by the equation $v_0 = k_m + (1 - a)k_8 + ak_9$. When CMB-treated HMM was used, the calculated values of a with 0.125 and 0.25 mg ml^{-1} F-actin were 0.93 and 0.75, respectively. They were in good accordance with the values of a measured directly by a light-scattering method and a Millipore filtration method, i.e. 0.91 and 0.72, respectively (From Inoue et al.[123], by courtesy of the Japanese Bio-chemical Society.)

the actomyosin–ATPase reaction occurs via one route [steps (1), (5), (6), (7) and (9)] and the recombination of F-actin with M_P^{ADP} is rate-determining.

By combining the mechanism for actomyosin–ATPase with that of the dissociation of actomyosin given above, the following mechanism is proposed for the myosin–actin–ATP system:

$$P_i$$
$$+$$
$$ADP \qquad\qquad\qquad A + M_1ATP \overset{(6)}{\rightleftharpoons} A + M_2ATP$$
$$+ \qquad\qquad (5) \nearrow \qquad\qquad\qquad\qquad \searrow (7)$$
$$AM + ATP \underset{(1)}{\rightleftharpoons} AM_1ATP \qquad\qquad\qquad\qquad A + M_P^{ADP}$$
$$(2) \searrow \qquad\qquad\qquad \nearrow (4) \qquad$$
$$AM_2ATP \underset{(3)}{\rightleftharpoons} AM_P^{ADP} \qquad (9)$$
$$\downarrow (8)$$

It should be noted that, in our previous monograph[34] and review[35], AM_P^{ADP} and $A + M_P^{ADP}$ in this scheme were denoted as the complex of F-actin with phosphoryl myosin, $AM_{\sim P}^{ADP}$, and F-actin + myosin–phosphate–ADP complex, $A + M_{A.,P}^{ADP}$, respectively, since the direct decomposition of AM_P^{ADP} was first suggested by us on the basis of analysis of the extra burst in the myosin–ATP system mentioned in Section 4.3.3. However, in this review $M_{\sim P}^{ADP}$ and $M_{A.,P}^{ADP}$ are not distinguished to simplify the mechanism.

In the above mechanism the dissociation of actomyosin is induced by steps (4) and (5), and the recombination of F-actin with myosin occurs at step (9). These two reactions correspond to detachment of the projections of myosin filaments from actin filaments and their attachment in the three fundamental steps in energy transduction of muscle described in Section 4.1. However, it has not yet been established in what step movement of the heads of the myosin molecule occurs. It seems probable that steps (5), (6) and (7) are not required for movement of the heads, since contraction of isolated sarcomeres could be induced by the addition of low concentrations of Mg^{2+}–$ATP^{125, 126}$. We[83] showed that actomyosin reconstituted from myosin modified with p-nitrothiophenol in the presence of ATP and a low concentration of Mg^{2+} ions* did not show step (8) or superprecipitation induced by ATP, although it had fairly high actomyosin–ATPase activity. Since it is impossible to reconstitute the myofibrillar structure from separated myosin and actin, superprecipitation of actomyosin has been used frequently as a model for muscle contraction in a disorganised state[2]. However, there are many problems in regarding superprecipitation of actomyosin as a contraction model, as described previously[34]. But if we disregard these problems, the above result suggests that movement of the myosin heads occurs in step (8). This suggestion is consistent with results obtained on transport ATPase[127, 128], showing that steps (3) and (8) are coupled with rotatory movement of the ATPase molecule, as described in Section 4.5.2.

---

*This p-nitrothiophenyl myosin had different properties from those of myosin modified with p-nitrothiophenol in the presence of ATP and a high concentration of Mg^{2+} ions.

In the presence of high concentrations of Mg^{2+}–ATP and the relaxing protein (a complex of tropomyosin with troponin), myofibrils relax when Ca^{2+} ions around myofibrils are removed by the sarcoplasmic reticulum[32,33]. X-Ray diffraction studies and measurements of elasticity of muscle fibres have indicated that in relaxed muscle fibres the heads of the myosin molecule are completely detached from F-actin filaments[34], as in the actomyosin–ATP system in the presence of a low concentration of actin and in the state of substrate inhibition. However, in relaxed muscle fibres the ATPase activity of actomyosin is decreased to the level of myosin-ATPase, while in the state of substrate inhibition of puried actomyosin the ATPase activity is higher than that of myosin. It is uncertain whether the nucleotide bound to myosin in relaxed muscle fibres is ATP or ADP. Bárány et al.[129] reported that the nucleotide bound to myosin in living muscle fibres at the relaxed state is ATP. This means that in the relaxed state myosin is trapped in a state of $A + M \cdot ATP$ by the binding of ATP to a regulatory site or that the state of $A + M_P^{ADP}$ is converted into $A + {}^{\circ}M \cdot ATP$ and stabilised by high concentrations of ATP. On the other hand, Marston and Treager[130,131] reported that the nucleotide bound to myosin in the relaxed state of glycerol-treated muscle fibres is ADP. If we accept this result, in the relaxed state the recombination of F-actin with M_P^{ADP} is inhibited by the conformational change in the actin–tropomyosin–troponin complex[132] or by steric hindrance due to migration of tropomyosin on the thin filament[133].

There are several problems in our measurements of the binding reaction of F-actin with HMM_P^{ADP}. The first is that the rate of binding of F-actin with HMM is very low (about $0.4\ s^{-1}$, cf. Figure 4.8), compared with the rate of muscle contraction. The second is that the rate of binding of F-actin with HMM is not given by $v = k[A][M_P^{ADP}]$. In other words, the rate is not linear with the F-actin concentration, especially at high concentrations of F-actin (Figure 4.8). The low rate of binding might be attributed to structural differences between myosin and HMM, since myosin has a filamentous structure at low ionic strength and the binding of F-actin with one myosin molecule may accelerate the rate of binding of neighbouring myosin molecules with F-actin. The rate of binding of myosin heads with F-actin in the actin–myosin–ATP system should be measured directly at low ionic strength, but this measurement might be very difficult.

The reason for the latter phenomenon is uncertain. It might be due to a change in conformation of myosin, such as the refractory state suggested by Eisenberg et al.[134] in which myosin cannot bind to F-actin. Although we obtained no indication of a refractory state in the actin–HMM–ATP system, our results clearly indicated that ITP induces a refractory state in HMM[135]. As shown in Figure 4.9, when ITP was added to acto-H-meromyosin the recombination of F-actin with HMM did not occur even when almost all the ITP had been decomposed and HMM had been freed from bound ITP, IDP and P_i. The rate of recombination of F-actin with HMM was about $1/6$ of that of liberation of IDP from HMM_P^{IDP}. However, addition of ITP to HMM alone did not induce the refractory state, and when F-actin was added at the time when almost all the nucleotide had been hydrolysed to IDP and P_i, rapid binding of F-actin with HMM was observed. Therefore it was concluded that the refractory state of myosin, in which myosin could not bind with F-actin, was induced not by the reaction of ITP with HMM, but by its reaction with acto-H-meromyosin:

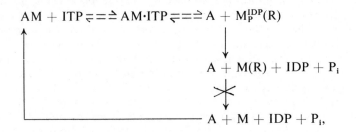

$$AM + ITP \rightleftharpoons AM \cdot ITP \rightleftharpoons A + M_P^{IDP}(R)$$

$$A + M(R) + IDP + P_i$$

$$A + M + IDP + P_i,$$

where the symbol (R) indicates the refractory state*.

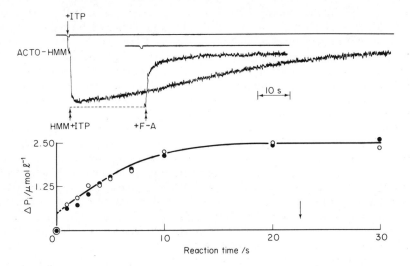

Figure 4.9 Changes in light-scattering intensity and liberation of P_i in the actin–H-meromyosin–ITP system. 0.17 mg ml^{-1} HMM, 0.1 mg ml^{-1} F-actin, 2.5 μM ITP, 50 mM KCl, 2 mM MgCl$_2$, 10 mM Tris-HCl, pH 7.8, 20 °C. The upper figure shows the time course of change in the light scattering intensity of acto-HMM on adding ITP (↓) and that when ITP was added to HMM alone (⋏) and then after completion of hydrolysis of ITP (see ↓ in the lower figure) F-actin was added (⋏). The reaction was followed in a stopped-flow apparatus. The upper two traces indicate flow rates. The lower figure shows the time course of liberation of P_i on adding ITP to HMM alone (○) or to acto-HMM (●). The HMM–ITPase reaction was not activated by F-actin at low concentrations of ITP. Its rate constant was equal to that of decomposition of HMM$_P^{IDP}$ measured from the recovery of change in the u.v. absorption spectrum, 0.6 s^{-1}. The rate constant of recovery of light-scattering intensity of acto-HMM after addition of ITP was about 0.1 s^{-1}, and was much lower than that of decomposition of HMM$_P^{IDP}$. When ITP was added to HMM alone and then F-actin was added after completion of hydrolysis of ITP, only the rapid increment of light-scattering intensity was observed

---

*M(R) is different from °M, since the former cannot bind with F-actin, whereas the latter can bind rapidly with F-actin, as mentioned above.

Finally, we will briefly mention the ATP–P_i exchange reaction in the actomyosin–ATPase reaction. As already mentioned on p. 139, in the case of myosin–ATPase reaction, scarcely any P exchange reaction occurred between ATP and M_P^{ADP} under usual experimental conditions. On the other hand, Hotta and Fujita[136] reported that P_i was incorporated into ATP during hydrolysis of ATP by actomyosin and that the amount of P_i incorporated into ATP was more than 10^{-5} of the amount of ATP hydrolysed by actomyosin. On the other hand, we[137] measured the ATP–P_i exchange reaction using purified actomyosin, but found no detectable incorporation of P_i into ATP, the amount being less than 10^{-3} of the amount of ATP hydrolysed. Recently, Wolcott and Boyer[138] reported that the amount of P_i incorporated into ATP was about 10^{-3} of that of ATP hydrolysed. The experimental conditions used by these workers were similar, so the reason for this discrepancy between the results is not clear. Further studies are needed on whether the ATP–P_i exchange reaction is directly related to the actomyosin–ATPase reaction or is catalysed by other proteins contaminating the actomyosin preparation.

4.5 ENERGY TRANSDUCING MECHANISM IN MUSCLE

4.5.1 Molecular mechanism of muscle contraction

As mentioned in Section 4.1, a basic molecular mechanism of muscle contraction has been provided by the sliding theory, and it is now well established that the development of tension and hydrolysis of ATP both result from the interaction between the projections from the thick filaments and the thin filaments. A. F. Huxley[139] assumed that there is spontaneous association and ATP-dependent dissociation between a contractile site, capable of oscillating for a certain distance along the back-bone of the thick filament, and the binding site on the actin filament. He showed that, given a simple form of probabilities of bonding and cleavage as functions of the position of the contractile site on the myosin filament, the sliding of the two filaments can adequately explain many of the mechanical and thermodynamic properties of contraction discovered by Hill[140]. This work became a starting point for many attempts to explain the mechanical and thermodynamic properties of contraction on the basis of a set of three fundamental reactions, i.e. the attachment of myosin heads to F-actin, the sliding of the two filaments by the movement of myosin heads, and the detachment of myosin heads from F-actin. Podolsky et al.[141, 142] later improved Huxley's model by analysing the change in length of muscle fibres after a rapid decrease in load. Furthermore, A. F. Huxley and Simmons[143, 144] measured the time course of the tension change after a sudden change in the length of the fibre, and proposed a new molecular kinetic model for muscle contraction. According to their hypothesis the angle formed by the myosin head and the actin filament determines the probability of binding of the myosin head with F-actin and their cleavage, and the sliding of the filaments is induced by rotation of the myosin head, which is bound with F-actin. These two analyses carried out by A. F. Huxley have become the basis of many models of contraction proposed later[145, 146].

As explained in detail in our monograph[34] and also mentioned in this review, there have been considerable advances in clarification of the reaction mechanisms of the three basic processes in contraction by biochemical studies on the reactions between myosin, actin and ATP. Molecular mechanisms in which contraction involves conformational changes of the myosin molecule on its reaction with ATP were presented by ourselves[147] and by Davies[148]. Transient kinetic studies on the myosin–actin–ATP system were first performed by us[34, 73, 75], and later by Taylor and co-workers[65, 76, 85]. They proposed an oversimplified variant of our original reaction mechanism of the system. Their mechanism has been used by many workers as a basis to propose a molecular mechanism of muscle contraction, but more recent studies on the reactions between myosin, actin and ATP have clearly indicated that our original mechanism is essentially correct, while Taylor's mechanism has many defects. However, as pointed out in each section of this review, there are several controversial problems to solve before a definite molecular model of muscle contraction based on biochemical results can be given. Firstly, it is uncertain whether the two heads of the myosin molecule are identical or not, and the physiological significance of the existence of two heads in the molecule is not clear. Secondly, it is uncertain how the two reaction processes of myosin–ATPase, discussed in Section 4.3, are related to the two heads of the myosin molecule, and what physiological function is served by the ATPase reaction via $^M_{ATP}$. Thirdly, it is uncertain how the two routes of actomyosin–ATPase are mutually related to each other. Fourthly, it is not known which step in the actomyosin–ATPase reaction is directly coupled with movement of cross bridges.

In Section 4.2 we presented evidence indicating that the two heads of the myosin molecule have different structures and functions from each other. However, as stated in Section 4.3.1, it is not clear whether the ATPase reactions of myosin via M_P^{ADP} and via $^M_{ATP}$ both take place in the same head or in different heads. Moreover, the physiological function of the ATPase reaction via $^M_{ATP}$ is not yet clear, although it has been established that the ATPase reaction via M_P^{ADP} is directly coupled with muscle contraction, as mentioned in detail in Section 4.4. The structural and functional interaction between the two heads of the myosin molecule is still uncertain, although the findings that actomyosin dissociates on adding 1 mole of ATP per 2 moles of the myosin head, while acto-HMM and acto-S-1 dissociate on adding 1 mole of ATP per mole of the myosin head, indicate a strong interaction between the two heads at least in the state where myosin binds to actin[104]. In connection with this, Shimizu and Yamada[149] have recently suggested that the two myosin heads, being functionally different and showing mutual interaction, are the functional basis for effective contraction. However, there is no direct experimental evidence for this, and further studies are required on the problem.

With regard to the third problem, we have obtained direct evidence for the two-route mechanism of actomyosin–ATPase, as described in Section 4.4, and the physiological meaning of each step of the two routes has been clarified fairly well. However, the mutual relationship between the two routes of the actomyosin–ATPase reaction is obscure. It seems probable that for a complete cycle of contraction to take place these two routes must occur alternately in the cycle, since one route of actomyosin–ATPase is for movement of the myosin heads, while the other is for detachment of the myosin heads from actin (*cf.* p. 148). However, this lacks direct experimental evidence. Moreover, as pointed out in Section 4.4.2, the rate of recombination of actin with HMM measured in the reconstituted system was very slow. The binding of the myosin heads with F-actin and their cleavage in an organised contractile system must be measured directly.

In connection with the fourth problem we showed from studies on myosin *p*-nitrothiophenylated in the presence of ATP and high concentrations of Mg^{2+}, and the one modified in the presence of ATP and low concentrations of Mg^{2+}, that both inhibition of formation of M_p^{ADP} and that of its direct decomposition suppress the superprecipitation of actomyosin on adding ATP, and concluded that the direct decomposition of AM_p^{ADP} is coupled with the movement of the myosin heads[34, 83]. However, this conclusion was deduced from studies on chemical modifications of myosin. It must be proved by more direct measurements of the movement of cross bridges, such as x-ray diffraction[150] and fluorescence polarisation studies[151] on the orientation of myosin heads in myofibrils on adding an ATP analogue, whose reaction with myosin stops at a certain step in the ATPase reaction.

Since there are still many problems to elucidate, we must submit a molecular mechanism of muscle contraction depending on reasonable assumptions. We[34, 74, 83] made the following assumptions. (i) Only one of the two heads in myosin forms M_p^{ADP}, while the other head does not react appreciably with ATP. (ii) S-1B, which forms M_p^{ADP} by reaction with ATP (*cf.* Section 4.2), can have at least two kinds of conformation, α and β, while the conformation of S-1A does not change during the contraction cycle. (iii) The conformation of S-1B, α or β, determines whether AM_p^{ADP} is directly decomposed into $AM + ADP + P_i$ or whether it dissociates into $A + M_p^{ADP}$. The direct decomposition of AM_p^{ADP} takes place when S-1B in the α-conformation reacts with ATP. S-1B is transformed by this reaction from the α- to the β-form. When S-1B in the β-state reacts with ATP, AM_p^{ADP} dissociates into $A + M_p^{ADP}$, and S-1B is converted into the original α-state. (iv) The binding of S-1A with F-actin is dependent on the conformation of S-1B, owing to steric hindrance or other effects. The binding of S-1A to the actin filament is weak when S-1B is in the α-conformation, but strong when S-1B is in the β-state.

We assume that in the relaxed state the myosin head S-1B is in the α-form and forms M_p^{ADP}, although it is controversial at present whether ATP or ADP binds to myosin in relaxed muscle fibres, as mentioned in Section 4.4. When Ca^{2+} ions are released from the sarcoplasmic reticulum by excitation of muscle fibres, these ions are received by troponin on the thin filaments. The binding of F-actin with M_p^{ADP} takes place owing to structural changes of the thin filaments induced by the Ca^{2+} binding[32, 132, 133]. As shown in Figure 4.10, F-actin binds with myosin at S-1B which is in the α-state. On the reaction of S-1B with ATP, AM_p^{ADP} is formed, and is decomposed directly into $AM + ADP + P_i$. Coupled with this reaction, S-1B is converted actively from the α- into the β-form, and sliding of the thin filaments past the thick filaments takes place. In the β-state of S-1B the two heads A and B both bind strongly with F-actin. The binding between S-1B and F-actin

154

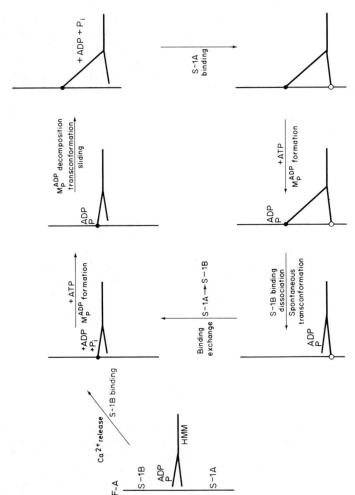

Figure 4.10 A molecular mechanism of muscle contraction. F-A represents F-actin. S-1A and S-1B are the two different heads of the myosin molecule. S-1B contains both the active site of the ATPase reaction and the site for strong binding with actin, while S-1A contains only a site for binding with F-actin, which is controlled by the conformation of S-1B. The conformation of S-1B undergoes a cyclic change of $\alpha \to \beta \to \alpha$, while the conformation of S-1A does not change during the contraction cycle. (From Hayashi and Tonomura[74], by courtesy of the Japanese Biochemical Society.)

is cleaved, when S-1B assumes the β-form, owing to the dissociation of AM_P^{ADP} into $A + M_P^{ADP}$, resulting in spontaneous reversion of S-1B to the α-conformation. Steric hindrance then ruptures the binding between S-1A and F-actin and the strong bond between S-1B and F-actin is reformed, so that the system returns to its original state. In this way the state of S-1B undergoes a cyclic transition $\alpha \rightarrow \beta \rightarrow \alpha$ with hydrolysis of two moles of ATP per mole of myosin. It was reported[152, 153] that the overall efficiency of energy transduction in muscle is 40–60%. Therefore it is necessary to assume that the energy liberated by ATP hydrolysis by the reaction via $A + M_P^{ADP}$ is also maintained in the myosin molecule in some form, and is used for development of tension.

4.5.2 Muscle contraction and active transport of cations

To conclude this review, the relation between the molecular mechanism of muscle contraction and those of other energy transductions will be mentioned briefly. In particular, the relationship between the molecular mechanisms of cation transport and muscle contraction is very interesting. Biological transport of solutes usually involves three major steps: recognition, translocation and release. Recognition must occur at the membrane boundary. Translocation through the membrane and release at the other side complete the transport process. Thus, the mechanism of transport appears to be very similar to the sliding mechanism of muscle contraction, which involves the binding of F-actin by myosin, translocation or sliding of F-actin coupled with ATP-splitting, and then release or dissociation of F-actin from myosin*.

The fragmented sarcoplasmic reticulum, SR, is very useful material for studies of the molecular mechanism of energy transduction, since its content of the Ca^{2+},Mg^{2+}-dependent ATPase is very high ($\sim 70\%$), and the coupling of Ca^{2+} uptake with ATP hydrolysis can be easily measured[34, 154]. As a result of much work on the SR-ATPase, the model shown in Figure 4.11 was presented for the coupling mechanism of ATP hydrolysis with Ca^{2+} uptake in the SR[34, 155]. In state E the enzyme binds strongly with Ca^{2+} ions, but not with Mg^{2+} ions[156]. One mole of ATP and 2 moles of Ca^{2+} ions bind from the outside of the membrane with enzyme E in a random sequence to form $^{Ca_2}E_1ATP$. The next complex, $^{Ca_2}E_2ATP$, is formed via $^{Ca_2}E_1ATP$. The complex $^{Ca_2}E_2ATP$ is transformed into the phosphorylated intermediate[157, 158], $^{Ca_2}E{\sim}P$, with release of ADP to the outside of the membrane. This reaction is coupled with translocation of Ca^{2+} ions from the outside to the inside of the membrane. Now the enzyme in state of $E{\sim}P$ can bind strongly with both Ca^{2+} and Mg^{2+} ions[156]. Therefore the intermediate, $^{Ca_2}E{\sim}P$, is converted into $^{Mg}E{\sim}P$ by replacement of Ca^{2+} ions on $E{\sim}P$ by Mg^{2+} ions, and the $^{Mg}E{\sim}P$ is hydrolysed. Both P_i and Mg^{2+} ions are liberated to the outside of the membrane.

Direct evidence for the translocation of Ca^{2+} ions through the membrane coupled with formation of $E{\sim}P$ from E_2ATP has recently been provided

---

*Cf. chapter 13 of Ref. 34 for details of similarities and differences between the reaction mechanisms of transport ATPase and contractile ATPase.

by measurement of Ca^{2+} uptake during each step in the Ca^{2+}, Mg^{2+}-dependent ATPase reaction[127]. Moreover, our studies[128] on the e.p.r. spectrum of spin-labelled SR suggested that rotation of the ATPase molecule is coupled with Ca^{2+} translocation. By measuring the time course of reduction of NEM spin-label bound to the SR on adding ascorbate, the percentages of spin label bound on the inside and the outside of the membrane were calculated. It was found that the ratio of spin labels bound on the inside and outside of the membrane changed markedly with formation of the reaction intermediates. The results obtained were all in accord with predictions from the model given in Figure 4.11, which was proposed to explain the kinetic properties of the coupling of Ca^{2+} transport and ATP hydrolysis, as mentioned above.

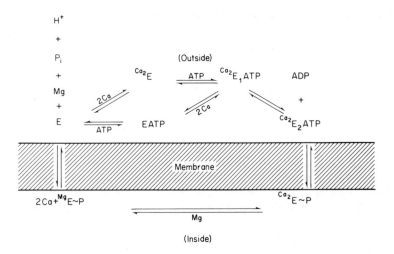

Figure 4.11 Mechanism of coupling of ATP hydrolysis with cation transport across the SR membrane. For details see text (From Tonomura[34], by courtesy of the University of Tokyo Press.)

Thus it is no exaggeration to say that the mechanism of coupling of the movement of the ATPase molecule with the physiological function in the SR membrane is better known than any other energy transduction, including muscle contraction. In studies on Ca^{2+} transport by the SR, the working hypothesis that the molecular mechanisms of muscle contraction and of active transport of cations are essentially the same has been very valuable for development of new ideas and methods. *Both in the transport and the contractile ATPase reaction, the key intermediate, $E{\sim}P$ or M_P^{ADP}, is formed via two kinds of enzyme–ATP complex[155,159], of which the first enzyme–ATP complex is in rapid equilibrium with enzyme + ATP, and the key reactions in muscle contraction and active transport of cations are movement of the ATPase molecule coupled with formation and decomposition of the key intermediate.* At first it was thought that the most evident difference between transport-ATPase and myosin–ATPase is the mode of binding of phosphate to enzyme in the key intermediate. Phosphate in the key intermediate of transport-ATPase binds covalently to a specific aspartic acid residue of enzyme[103, 160], while phosphate in the key intermediate of myosin–ATPase is liberated as P_i on adding TCA. However, as already stated in Section 4.3.3, we showed that the values of $\Delta H°$ and $\Delta S°$ of the step $M_2ATP \rightleftharpoons M_P^{ADP}$ in the myosin–ATPase reaction are equal to those of the reaction between E_2ATP and $E_{\sim P}^{ADP}$ in the Na^+,K^+-dependent ATPase reaction. This result suggested a similarity in the chemical structures of the key intermediates in the two ATPase reactions.

The hypothesis that the molecular mechanisms of contraction and active transport are essentially the same has been proved to be a very useful working hypothesis, and we can expect that results in one field can be used to solve some problems in the other, resulting in mutual progress in both fields. Furthermore, since it has recently been established that the Na^+,K^+-dependent ATPase[161, 162] and the Ca^{2+},Mg^{2+}-dependent ATPase reactions[163-166] are completely reversible, studies on transport-ATPase should also be valuable for elucidation of the molecular mechanism of ATP synthesis in the membrane transducing biological energy. Thus, we hope that in the near future studies at the molecular level on muscle contraction and active transport of cations will enable us to elucidate the molecular basis, and hence establish the basic principles, of biological energy transductions in general.

Acknowledgement

The work described in this review was supported by grants from the Ministry of Education of Japan and the Muscular Dystrophy Associations of America, Inc.

Note added in Proof

Recently, we [Inoue, A. and Tonomura, Y. (1975). *J. Biochem.*, in press] obtained evidence for the involvement of $_{ATP}^M$ in the regulation of actomyosin–ATPase by trace amounts of Ca^{2+} ions: in the absence of Ca^{2+} ions where the actin-relaxing protein–HMM complex shows the myosin type of ATPase activity, 1 mole of HMM moiety contains 1 mole of ATP bound to the site for route (4.2), and HMM is in a mixed state of M_P^{ADP} and $°M$ for the reaction of route (4.1). In the presence of a minute amount of Ca^{2+} ions, where the protein complex shows a high ATPase activity of the actomyosin type, HMM is in the state of M_P^{ADP} without binding of ATP at the site for route (4.2). Thus, the Ca^{2+}-sensitive dissociation of actomyosin in the presence of the relaxing protein is induced by the formation of $_{ATP}^M$ and this dissociation corresponds to the relaxation of muscle, while the Ca^{2+}-insensitive dissociation due to the formation of M_P^{ADP} is required for the sliding of the actin filaments past the myosin filaments, as mentioned in Section 4.5.1.

References

1. Engelhardt, V. A. (1946). *Adv. Enzymol.*, Vol. 6, 147 (N. N. Nord, editor) (New York: Interscience)
2. Szent-Györgyi, A. (1947 & 1951). *Chemistry of Muscular Contraction*, 1st ed., and 2nd ed. (New York: Academic Press)
3. Weber, H. H. and Portzehl, H. (1954). *Progr. Biophys. Biophys. Chem.*, Vol. 4, 60 (J. A. V. Butler and J. T. Randall, editors) (London: Pergamon)
4. Perry, S. V. and Corsi, A. (1958). *Biochem. J.*, **68,** 5
5. Portzehl, H. (1951). *Z. Naturforsch.*, **6b,** 355
6. Hayashi, T. (1952). *J. Gen. Physiol.*, **36,** 139
7. Cain, D. F. and Davies, R. E. (1962). *Biochem. Biophys. Res. Commun.*, **8,** 361
8. Huxley, A. F. and Niedergerke, R. (1954). *Nature*, **173,** 971
9. Huxley, H. E. and Hanson, J. (1954). *Nature*, **173,** 973
10. Huxley, H. E. (1953). *Biochim. Biophys. Acta*, **12,** 387
11. Huxley, H. E. (1957). *J. Biochem. Biophys. Cyt.*, **3,** 631
12. Hanson, J. and Huxley, H. E. (1953). *Nature*, **172,** 530
13. Hasselbach, W. (1953). *Z. Naturforsch.*, **8b,** 449
14. Szent-Györgyi, A. G. (1951). *Arch. Biochem. Biophys.*, **31,** 97
15. Hanson, J. and Lowy, J. (1963). *J. Mol. Biol.*, **6,** 46
16. Ramsey, R. W. and Street, S. F. (1940). *J. Cell. Comp. Physiol.*, **15,** 11
17. Gordon, A. M., Huxley, A. F. and Julian, F. J. (1966). *J. Physiol.*, **184,** 170
18. Sandberg, J. A. and Carlson, F. D. (1966). *Biochem. Z.*, **345,** 212
19. Ward, P. C. J., Edwards, C. and Benson, E. S. (1965). *Proc. Nat. Acad. Sci. USA*, **53,** 1377
20. Hayashi, Y. and Tonomura, Y. (1968). *J. Biochem.*, **63,** 101
21. Fukazawa, T., Hashimoto, Y. and Tonomura, Y. (1963). *Biochim. Biophys. Acta*, **75,** 234
22. Huxley, H. E. and Brown, W. (1967). *J. Mol. Biol.*, **30,** 387
23. Huxley, H. E. (1963). *J. Mol. Biol.*, **7,** 281
24. Lowey, S., Slayter, H. S., Weeds, A. G. and Baker, H. (1969). *J. Mol. Biol.*, **42,** 1
25. Heilbrunn, L. V. (1940). *Physiol. Zool.*, **13,** 88
26. Heilbrunn, L. V. and Wircinsky, F. J. (1947). *J. Cell. Comp. Physiol.*, **29,** 15
27. Kamada, K. and Kinoshita, H. (1943). *Jap. J. Zool.*, **10,** 469
28. Bozler, E. (1954). *J. Gen. Physiol.*, **38,** 149
29. Watanabe, S. (1955). *Arch. Biochem. Biophys.*, **54,** 559
30. Hasselbach, W. and Makinose, M. (1961). *Biochem. Z.*, **333,** 518
31. Ebashi, S. and Lipmann, F. (1962). *J. Cell. Biol.*, **14,** 389
32. Ebashi, S. and Endo, M. (1968). *Progr. Biophys. Mol. Biol.*, Vol. 18, 123 (J. A. V. Butler and D. Noble, editors) (London: Pergamon)
33. Weber, A. and Murray, J. M. (1973). *Physiol. Rev.*, **53,** 612
34. Tonomura, Y. (1972). *Muscle Proteins, Muscle Contraction and Cation Transport* (Tokyo and Baltimore: Univ. Tokyo Press and Univ. Park Press)
35. Tonomura, Y. and Inoue, A. (1974). *Mol. Cell. Biochem.*, **5,** 127
36. Tonomura, Y. and Oosawa, F. (1972). *Ann. Rev. Biophys.*, **1,** 159
37. Perry, S. V. (1967). *Progr. Biophys. Mol. Biol.*, Vol. 17, 325 (J. A. V. Butler and H. E. Huxley, editors) (London: Pergamon)
38. Slayter, H. S. and Lowey, S. (1967). *Proc. Nat. Acad. Sci. USA*, **58,** 1611
39. Szent-Györgyi, A. G. (1953). *Arch. Biochem. Biophys.*, **42,** 305
40. Mueller, H. and Perry, S. V. (1962). *Biochem. J.*, **85,** 431
41. Lowey, S., Goldstein, L., Cohen, C. and Luck, S. M. (1967). *J. Mol. Biol.*, **23,** 287
42. Botts, J., Cook, R., dos Remedios, C., Duke, J., Mendelson, R., Morales, M. F., Tokiwa, T., Veniegra, G. and Yount, R. (1973). *Cold Spring Harbor Symp. Quant. Biol.*, **37,** 195
43. Starr, R. and Offer, G. W. (1973). *J. Mol. Biol.*, **81,** 17
44. Tsao, T.-C. (1953). *Biochim. Biophys. Acta*, **11,** 368
45. Kielley, W. W. and Harrington, W. F. (1960). *Biochim. Biophys. Acta*, **41,** 401
46. Frederiksen, D. W. and Holtzer, A. (1968). *Biochemistry*, **7,** 3935
47. Sarker, S. and Cooke, P. H. (1970). *Biochem. Biophys. Res. Commun.*, **41,** 918

48. Dow, J. and Stracher, A. (1971). *Proc. Nat. Acad. Sci. USA*, **68**, 1107
49. Offer, G. W. (1965). *Biochim. Biophys. Acta*, **111**, 191
50. Yazawa, M. and Yagi, K. (1974). Personal communication
51. Weeds, A. G. and Lowey, S. (1971). *J. Mol. Biol.*, **61**, 701
52. Hayashi, Y. (1972). *J. Biochem.*, **72**, 82
53. Stracher, A. (1969). *Biochem. Biophys. Res. Commun.*, **35**, 519
54. Dreizen, P. and Gershman, L. C. (1970). *Biochemistry*, **9**, 1688
55. Lowey, S. and Risby, D. (1971). *Nature*, **234**, 81
56. Weeds, A. G. and Franc, G. (1973). *Cold Spring Harbor Symp. Quant. Biol.*, **37**, 9
57. Hayashi, Y., Takenaka, H. and Tonomura, Y. (1973). *J. Biochem.*, **74**, 1031
58. Tonomura, Y. and Morita, F. (1959). *J. Biochem.*, **46**, 1367
59. Nauss, K. M., Kitagawa, S. and Gergely, J. (1969). *J. Biol. Chem.*, **244**, 755
60. Young, D. M. (1967). *J. Biol. Chem.*, **242**, 2790
61. Morita, F. (1971). *J. Biochem.*, **69**, 513
62. Yazawa, M., Morita, F. and Yagi, K. (1973). *J. Biochem.*, **74**, 1107
63. Ouellet, L., Laidler, K. J. and Morales, M. F. (1952). *Arch. Biochem. Biophys.*, **39**, 37
64. Watanabe, S., Tonomura, Y. and Shiokawa, H. (1953). *J. Biochem.*, **40**, 387
65. Lymn, R. W. and Taylor, B. W. (1970). *Biochemistry*, **9**, 2975
66. Inoue, A., Shibata-Sekiya, K. and Tonomura, Y. (1972). *J. Biochem.*, **71**, 115
67. Bowen, W. J. and Evans, J. C. Jr. (1968). *Europ. J. Biochem.*, **5**, 507
68. Schliselfeld, L. H. and Bárány, M. (1968). *Biochemistry*, **7**, 3206
69. Inoue, A. and Tonomura, Y. (1974). *J. Biochem.*, **76**, 755
70. Kinoshita, N., Kubo, S., Onishi, H. and Tonomura, Y. (1969). *J. Biochem.*, **65**, 285
71. Yagi, K., Yazawa, Y., Ohtani, F. and Okamoto, Y. (1972). Presented at Japan–U.S. Seminar, Tokyo
72. Tonomura, Y. and Kitagawa, S. (1960). *Biochim. Biophys. Acta*, **40**, 135
73. Kanazawa, T. and Tonomura, Y. (1965). *J. Biochem.*, **57**, 604
74. Hayashi, Y. and Tonomura, Y. (1970). *J. Biochem.*, **68**, 665
75. Tokiwa, T. and Tonomura, Y. (1965). *J. Biochem.*, **57**, 616
76. Finlayson, B. and Taylor, E. W. (1970). *Biochemistry*, **8**, 802
77. Green, I. and Mommaerts, W. F. H. M. (1953). *J. Biol. Chem.*, **202**, 541
78. Imamura, K., Kanazawa, T., Tada, M. and Tonomura, Y. (1965). *J. Biochem.*, **57**, 627
79. Tonomura, Y., Kitagawa, S. and Yoshimura, J. (1962). *J. Biol. Chem.*, **237**, 3660
80. Morita, F. and Yagi, K. (1966). *Biochem. Biophys. Res. Commun.*, **22**, 297
81. Morita, F. (1967). *J. Biol. Chem.*, **242**, 4501
82. Sekiya, K. and Tonomura, Y. (1967). *J. Biochem.*, **61**, 787
83. Tonomura, Y., Nakamura, H., Kinoshita, N., Onishi, H. and Shigekawa, M. (1969). *J. Biochem.*, **66**, 599
84. Inoue, A. and Tonomura, Y. (1973). *J. Biochem.*, **73**, 555
85. Taylor, E. W., Lymn, R. W. and Moll, G. (1970). *Biochemistry*, **9**, 2984
86. Nakamura, H. and Tonomura, Y. (1968). *J. Biochem.*, **63**, 279
87. Malik, M. N. and Martonosi, A. (1972). *Arch. Biochem. Biophys.*, **152**, 243
88. Martonosi, A. and Malik, M. N. (1973). *Cold Spring Harbor Symp. Quant. Biol.*, **37**, 184
89. Werber, M. M., Szent-Györgyi, A. G. and Fasman, G. D. (1972). *Biochemistry*, **11**, 2872
90. Morita, F. (1972). *Molecular Mechanism of Enzyme Action*, 282 (Y. Ogura, Y. Tonomura and T. Nakamura, editors) (Tokyo: Univ. Tokyo Press)
91. Seidel, J. C. and Gergely, J. (1971). *Biochem. Biophys. Res. Commun.*, **44**, 826
92. Seidel, J. C. and Gergely, J. (1973). *Cold Spring Harbor Symp. Quant. Biol.*, **37**, 187
93. Seidel, J. C. and Gergely, J. (1973). *Arch. Biochem. Biophys.*, **158**, 853
94. Trentham, D. R., Bardsley, R. G., Eccleston, J. E. and Weeds, A. G. (1970). *Biochem. J.*, **126**, 635
95. Arata, T., Inoue, A. and Tonomura, Y. (1974). *J. Biochem.*, **76**, 1211
96. Onishi, H., Ohtsuka, E., Ikehara, M. and Tonomura, Y. (1973). *J. Biochem.*, **74**, 435
97. Wolcott, R. G. and Boyer, P. D. (1972). *Biochem. Biophys. Res. Commun.*, **51**, 428
98. Sartorelli, L., Fromm, H. J., Benson, R. W. and Boyer, P. D. (1966). *Biochemistry*, **11**, 2877
99. Bagshaw, C. R. and Trentham, D. R. (1973). *Biochem. J.*, **133**, 323
100. Inoue, A., Arata, T. and Tonomura, Y. (1974). *J. Biochem.*, **76**, 661

101. Fukushima, Y. and Tonomura, Y. (1975). *J. Biochem.*, **77**, 533
102. Nagano, K., Kanazawa, T., Mizuno, N., Tashima, Y., Nakao, T. and Nakao, M. (1965). *Biochem. Biophys. Res. Commun.*, **19**, 759
103. Bastide, F., Meissner, G., Fleischer, S. and Post, R. L. (1973). *J. Biol. Chem.*, **248**, 8385
104. Takeuchi, K. and Tonomura, Y. (1969). *J. Biochem.*, **70**, 1011
105. Tregear, R. T. and Squire, J. M. (1973). *J. Mol. Biol.*, **77**, 279
106. Podolsky, R. J. and Teichholz, L. E. (1970). *J. Physiol.*, **211**, 19
107. Onishi, H., Nakamura, H. and Tonomura, Y. (1968). *J. Biochem.*, **64**, 769
108. Morita, F. and Tonomura, Y. (1960). *J. Amer. Chem. Soc.*, **82**, 5172
109. Kinoshita, N., Kanazawa, T., Onishi, H. and Tonomura, Y. (1969). *J. Biochem.*, **65**, 567
110. Lymn, R. W. and Taylor, E. W. (1971). *Biochemistry*, **10**, 4617
111. Tonomura, Y. and Kanazawa, T. (1965). *J. Biol. Chem.*, **240**, CP 4110
112. Sekiya-Shibata, K. and Tonomura, Y. (1975). *J. Biochem.*, **77**, 543
113. Onishi, H., Nakamura, H. and Tonomura, Y. (1968). *J. Biochem.*, **63**, 739
114. Tonomura, Y. and Yoshimura, J. (1960). *Arch. Biochem. Biophys.*, **90**, 73
115. Straub, F. B. and Feuer, G. (1950). *Biochim. Biophys. Acta*, **4**, 455
116. Mommaerts, W. F. H. M. (1951). *Biochim. Biophys. Acta*, **7**, 477
117. Asakura, S., Taniguchi, M. and Oosawa, F. (1963). *J. Mol. Biol.*, **7**, 55
118. Szent-Györgyi, A. G. and Prior, G. (1966). *J. Mol. Biol.*, **15**, 515
119. Nakaoka, Y. and Kasai, M. (1969). *J. Mol. Biol.*, **44**, 319
120. Moos, C. and Eisenberg, E. (1970). *Biochim. Biophys. Acta*, **223**, 221
121. Bárány, M., Tucci, A. F. and Conover, T. E. (1966). *J. Mol. Biol.*, **19**, 483
122. Tokiwa, T., Shimada, T. and Tonomura, Y. (1967). *J. Biochem.*, **61**, 108
123. Inoue, A., Shigekawa, M. and Tonomura, Y. (1973). *J. Biochem.*, **74**, 923
124. Leadbeater, L. and Perry, S. V. (1963). *Biochem. J.*, **87**, 233
125. Takahashi, K., Mori, T., Nakamura, H. and Tonomura, Y. (1965). *J. Biochem.*, **57**, 637
126. Nakamura, H., Mori, T. and Tonomura, Y. (1965). *J. Biochem.*, **58**, 582
127. Sumida, M. and Tonomura, Y. (1974). *J. Biochem.*, **75**, 283
128. Tonomura, Y. and Morales, M. F. (1974). *Proc. Nat. Acad. Sci. USA*, **71**, 3687
129. Bárány, M. and Bárány, K. (1973). *Cold Spring Harbor Symp. Quant. Biol.*, **37**, 157
130. Marston, S. B. and Tregear, R. T. (1972). *Nature New Biol.*, **235**, 23
131. Marston, S. B. (1973). *Biochim. Biophys. Acta*, **305**, 397
132. Tonomura, Y., Watanabe, S. and Morales, M. F. (1969). *Biochemistry*, **8**, 2171
133. Huxley, H. E. (1973). *Cold Spring Harbor Symp. Quant. Biol.*, **37**, 361
134. Eisenberg, E., Dobkin, L. and Kielley, W. W. (1972). *Proc. Nat. Acad. Sci. USA*, **69**, 667
135. Inoue, A., Tonomura, Y. and Watanabe, S. (1975). *J. Biochem.*, in the press
136. Hotta, K. and Fujita, Y. (1971). *Physiol. Chem. Phys.*, **3**, 196
137. Inoue, A. (1973). *J. Biochem.*, **73**, 1311
138. Wolcott, R. G. and Boyer, P. D. (1974). *Biochem. Biophys. Res. Commun.*, **57**, 709
139. Huxley, A. F. (1957). *Progr. Biophys. Biophys. Chem.*, Vol. 7, 255 (J. A. V. Butler and B. Katz, editors) (London: Pergamon)
140. Hill, A. V. (1938). *Proc. Roy. Soc.*, **B126**, 136
141. Podolsky, R. J., Nolan, A. C. and Zaveler, S. A. (1969). *Proc. Nat. Acad. Sci. USA*, **64**, 504
142. Podolsky, R. J. and Nolan, A. C. (1973). *Cold Spring Harbor Symp. Quant. Biol.*, **37**, 661
143. Huxley, A. F. and Simmons, R. M. (1971). *Nature*, **233**, 533
144. Huxley, A. F. and Simmons, R. M. (1973). *Cold Spring Harbor Symp. Quant. Biol.*, **37**, 669
145. Hill, T. L. (1973). *Proc. Nat. Acad. Sci. USA*, **70**, 2732
146. White, D. C. S. and Thorson, J. (1973). *Progr. Biophys. Mol. Biol.*, Vol. 27, 173 (A. J. V. Butler and D. Noble, editors) (London: Pergamon)
147. Tonomura, Y., Yagi, K., Kubo, S. and Kitagawa, S. (1961). *J. Res. Inst. Catalysis, Hokkaido Univ.*, **9**, 256
148. Davies, D. E. (1963). *Nature*, **199**, 1068
149. Shimizu, H. and Yamada, M. (1975). *J. Theor. Biol.*, **49**, 89
150. Lymn, R. W. and Huxley, H. E. (1973). *Cold Spring Harbor Symp. Quant. Biol.*, **37**, 449

151. Dos Remedios, C. G., Yount, R. G. and Morales, M. F. (1972). *Proc. Nat. Acad. Sci. USA*, **69**, 2542
152. Kushmerick, M. J. and Davies, R. E. (1969). *Proc. Roy Soc.*, **B174**, 293
153. Pybus, J. and Tregear, R. T. (1973). *Cold Spring Harbor Symp. Quant. Biol.*, **37**, 655
154. Martonosi, A. (1972). *Current Topics in Membranes and Transport*, Vol. 3, 83 (F. Bronner and A. Kleinzeller, editors) (New York: Academic Press)
155. Kanazawa, T., Yamada, S., Yamamoto, T. and Tonomura, Y. (1971). *J. Biochem.*, **70**, 95
156. Yamada, S. and Tonomura, Y. (1972). *J. Biochem.*, **72**, 417
157. Yamamoto, T. and Tonomura, Y. (1967). *J. Biochem.*, **62**, 558
158. Makinose, M. (1969). *Europ. J. Biochem.*, **10**, 74
159. Kanazawa, T., Saito, M. and Tonomura, Y. (1970). *J. Biochem.*, **67**, 693
160. Degani, C. and Boyer, P. D. (1973). *J. Biol. Chem.*, **248**, 8222
161. Glynn, I. M. and Garrahan, P. J. (1972). *J. Physiol.*, **192**, 237
162. Post, R. L., Taniguchi, K. and Toda, G. (1974). *Ann. New York Acad. Sci.*, **242**, 80
163. Makinose, M. (1972). *FEBS Lett.*, **25**, 113
164. Yamada, S., Sumida, M. and Tonomura, Y. (1972). *J. Biochem.*, **72**, 1537
165. Kanazawa, T. and Boyer, P. D. (1973). *J. Biol. Chem.*, **248**, 3163
166. Kanazawa, T. (1975). *J. Biol. Chem.*, **250**, 113

5
Control of Energy Transducing Systems

E. RACKER
Cornell University

5.1 INTRODUCTION

While the mechanism of energy transduction is under extensive investigation in many laboratories all over the world, the control of energy transducing systems has received scanty attention. A reader of a current journal devoted to bioenergetic or membrane research might receive the erroneous impression that we either know all there is to know about energy control or that we know so little that we cannot examine it experimentally. Neither impression is correct, since numerous avenues of investigation are open. It is a universal feature of nature that energy-linked processes are 'tightly' coupled. The energy relationships are fixed and usually no more energy is expended than necessary. But there are examples that the tight coupling can be loosened so that extra heat is produced, for example in the brown fat pads of animals exposed to the cold or during involution of lactating mammary glands. There is a clinical case of a loosely coupled woman, or stated properly, a woman with loosely coupled mitochondria, who is incapacitated by the inefficiency of her muscles[1]. There are diseases in which the mitochondria, glycolytic enzymes or some energy-dependent transport mechanisms are defective and bring about alterations in function. The lesions may be in the energy transducing mechanism, e.g. in the sarcoplasmic reticulum in Dŭchenne muscular distrophy[2] or may be elsewhere, e.g. in fatty acid metabolism[3] and affect mitochondria indirectly. The role of fatty acids as natural uncouplers will be discussed later. These examples illustrate the old principle that pathology makes us aware of the normal state of affairs, which indeed would appear miraculous, were it not normal.

Most cells of animals, plants or bacteria have only one or two sources of energy income, but innumerable ways of spending it. It's like your budget, particularly if you are married. This state of affairs requires the installation of some controls. How is order established inside the cells involving hundreds of participating enzymes? What are the means of communication and restriction? The first degree or order is established by compartmentation with the help of membranes. The cell has a plasma membrane which represents a barrier for the export as well as the import of many solutes. Inside there are subcellular structures with highly specialised membranes that regulate the flow of ions and macromolecules. Communication between the compartments is usually established by relatively small molecular substances such as adenine nucleotides or carboxylic acids. Some membranes are endowed with specific transport systems to accomodate these movements. Compounds which interfere with these transport systems, like oligomycin, atractyloside, bongkrekic acid or mersalyl, are severe poisons on the one hand and most useful biochemical tools on the other.

The second degree of control is established by the intrinsic properties of the enzymes and multienzyme systems. Examples for both allosteric and allotopic properties of enzymes involved in energy metabolism will be discussed in greater detail later.

A third level of control is exerted by compounds that are designed for the specific purpose of regulation. These compounds may be proteins firmly attached to an enzyme and are called regulatory subunits or they may be dissociable proteins or smaller molecular compounds like cyclic AMP, cyclic GMP, prostaglandin or fatty acids. These small molecular substances may act directly or indirectly via a regulatory subunit of the enzyme.

These controls are operative both at the level of energy generation and utilisation and will be discussed under these headings. Yet it should be realised that one of the fundamental principles of energy control in nature is the close linkage between the utilisation of ATP (the ATPase activity in the broadest sense) and the generation of ATP in either glycolysis or oxidative phosphorylation or photophosphorylation. In all three systems, the oxidation step responsible for ATP generation is tightly coupled, i.e. oxidation ceases when ADP and P_i are not available.

5.2   CONTROL OF ENERGY GENERATION

5.2.1   Glycolysis and the Pasteur effect

The most important regulatory phenomenon of glycolysis, known as the Pasteur effect, was recognised long before the fundamentals of fermentation were known. Thus it is not surprising that early speculations on the mechanism of this regulatory phenomenon were far off the mark. What Pasteur observed was that under aerobic conditions yeast cells consumed less glucose per unit weight than under anaerobic conditions. The first important clue to this phenomenon and to the elucidation of the Pasteur effect was the observation that nitrophenols eliminate the inhibition of glycolysis[4]. Since these compounds did not inhibit respiration, oxygen itself, which was featured in earlier hypotheses, could no longer be implicated directly in the Pasteur effect.

When it was recognised[5] that the nitrophenols are uncouplers of oxidative phosphorylation it became apparent that the inhibition of glucose utilisation was dependent on the production of ATP by mitochondria and that the Pasteur effect is a phenomenon of bioenergetics control. There are few subjects in biochemistry more confusing to students than the Pasteur effect, mainly because of the abundance of unfounded speculations. If investigators had concentrated on the basic phenomenon observed by Pasteur, namely the inhibition of sugar utilisation under aerobic conditions, we would have been spared a lot of confusion. However, because brilliant and convenient manometric methods for the measurements of the end-products of fermentation, i.e. lactic acid or CO_2, had been developed by Warburg, analysis as well as thinking concentrated on the end rather than on the beginning of the glucose pathway. The lesson which we may learn from this mistake is that the thought is more important than the medium and that we should resist the temptation of being seduced by instruments.

As we see it now, the Pasteur effect is a complex and coordinated system of controls which operate at least at three levels: (a) at the step of glyceraldehyde 3-phosphate oxidation, and at the steps of phosphorylation of (b) glucose and (c) fructose 6-phosphate.

5.2.1.1   *Control of glyceraldehyde 3-phosphate oxidation*

Glyceraldehyde 3-phosphate, which is derived by the cleavage of phosphorylated sugar, is oxidised in the presence of glyceraldehyde 3-phosphate dehydrogenase and NAD to yield a phosphoglyceryl-enzyme and NADH.

Before the next step, the phosphorolysis of the acyl enzyme and the formation of 1,3-diphosphoglycerate, can take place, NADH must be removed because it is a potent allosteric inhibitor of the phosphorolytic (or hydrolytic) removal of the acyl group[6]. That the inhibition is allosteric, rather than a reversal of oxidation, is apparent from two observations: NADH inhibits in the presence of arsenate, which ensures irreversibility of the oxidation step. Secondly, gentle proteolytic digestion of the enzyme eliminates the inhibitory effect of NADH, while oxidation is unimpaired[6]. In the normal course of glycolysis, NADH is removed and NAD is regenerated when pyruvate is reduced by lactate dehydrogenase. NAD then recombines with glyceraldehyde 3-phosphate dehydrogenase and renders it susceptible to the next step of phosphorolysis. Thus there is a compulsory oxido-reduction step prior to the entry of P_i. The product of phosphorolysis, 1,3-diphosphoglycerate, is also an inhibitor of glyceraldehyde 3-phosphate oxidation and is removed by ADP to yield ATP. Thus the progress of oxidation is dependent on the availability of ADP and is coupled to a phosphorylation step. This multiple mutual dependency of oxido-reduction and phosphorylation insures tight coupling. Uncoupling can take place at the level of the acyl enzyme or of 1,3-diphosphoglycerate. Arsenate disposes of the high-energy acyl intermediate on the enzyme by substituting for inorganic phosphate, while an acyl phosphatase present in tissues[7] hydrolyses 1,3-diphosphoglycerate. A curious phenomenon of uncoupling of glyceraldehyde 3-phosphate oxidation has recently been reported[8]. Several uncouplers of oxidative phosphorylation were reported to inhibit phosphorylation ($^{32}P_i$–ATP exchange) associated with glyceraldehyde 3-phosphate dehydrogenase. However, rather high concentrations of uncouplers were required. The possibility that these compounds, which are proton ionophores, may uncouple by entering a hydrophobic active site of the enzyme should be considered. Actually, in spite of extensive investigations of this crystalline enzyme the fate of the proton which should be released during the reaction of NAD with an SH group of the enzyme[9] is still unknown. Although these observations are of interest from a theoretical point of view it is quite clear that these uncouplers do not affect glyceraldehyde 3-phosphate dehydrogenase in intact cells, since glycolytic ATP formation is maintained or increased in the presence of uncouplers while mitochondrial ATP formation is eliminated[10].

Physiologically, the most important release of oxidation control in glycolysis is by hydrolysis of ATP. We shall refer to this process under the name of 'ATPase activity' in the broadest sense, which includes all work processes such as ion translocation and biosynthesis of macromolecules. We shall discuss later the large variety of metabolic activities that contribute to the regeneration of ADP and P_i by this 'ATPase activity'.

5.2.1.2 Control of glucose and fructose 6-phosphate phosphorylation

Little would be gained by a control of glyceraldehyde 3-phosphate oxidation if the utilisation of glucose remained unimpaired. In fact the cell would go into ionic inbalance by depletion of inorganic phosphate owing to accumulation of hexose phosphates. Such an energy crisis would undoubtedly be fatal.

Attempts have actually been made to induce tumor cells to commit suicide by such a mechanism by feeding them deoxyglucose. Although tumor cells have usually higher hexokinase activities than most other cells and thus more likely to commit such a suicide, the tolerance range of toxicity of deoxyglucose is too narrow for comfort. Yet for bacteria with entirely different fermentation potentials one might succeed in designing a 'sweet death' with such a suicidal sugar as a chemotherapeutic agent.

Restriction of glucose uptake is therefore an essential feature of the Pasteur effect and is accomplished by the coordinated allosteric control of phosphofructokinase and hexokinase. It was shown by Lardy and Park[11] that the phosphorylation of fructose 6-phosphate was inhibited by ATP. Later, several secondary effectors, including inorganic phosphate, were found that counteract the inhibition of ATP[12]. An inhibited phosphofructo-kinase leads to an increase in intracellular concentration of glucose 6-phosphate, which in turn inhibits glucose phosphorylation by hexokinase. The latter inhibition is also counteracted by inorganic phosphate[13, 14].

We mention specifically the effect of phosphate because in ascites tumor cells[10], yeast cells[15] and muscle[16] the intracellular levels of ATP do not change significantly during transition from anaerobic to aerobic conditions while that of secondary allosteric effectors such as inorganic phosphate or phosphocreatine do. It is therefore apparent that the Pasteur effect is a complex phenomenon not only because its controls operate at several levels of the glycolytic pathway but because a multitude of different secondary allosteric effectors act on the same step, thus allowing for a range of variations of metabolic control depending on the physiological needs of different cells.

The above mentioned principles participating in the operation of a Pasteur effect have been demonstrated in reconstituted systems[17], including the requirement for allosterically controlled hexokinase and phosphofructokinase enzymes. For example, yeast hexokinase which is not allosterically controlled by glucose 6-phosphate could not be substituted for the glucose 6-phosphate-sensitive mammalian enzyme. This of course raises the interesting question how the Pasteur effect, which was first discovered in yeast, operates in these micro-organisms. Unpublished experiments in our laboratory suggest that mannose 6-phosphate, a metabolic side product, may serve in place of glucose 6-phosphate as an allosteric effector in yeast. This subtle variation on the theme of the Pasteur effect may have physiological significance in cells that require relatively high concentrations of glucose 6-phosphate for the operation of an active pentose phosphate cycle.

Lynen and Königsberger[18] suggested many years ago that ATP generated in mitochondria might not be available for hexokinase. This possibility of compartmentation as a component of the Pasteur effect was recently revived by Rapoport and collaborators[19], who have found that in reticulocytes (a) the phosphofructokinase control is insufficient to explain the Pasteur effect and (b) that adenine nucleotide compartmentation in the mitochondria contributes to the restriction of glucose phosphorylation. Biosynthesis and degradation of adenine nucleotides may also control glycolysis[20].

Conceivably, additional and completely different mechanisms may operate in the Pasteur effect. The reviewer pointed out over 20 years ago[21] that several alternative mechanisms may have been selected during evolution for the control of glucose utilisation, thereby helping cells to live more economically for longer periods of time. The reviewer therefore suggested a search for multiple 'Pasteur inhibitors' and although many have since been found[17], we have not reached the end of the line.

5.2.2 Oxidative phosphorylation

5.2.2.1 *Control of the respiratory chain*

The mechanism of oxidative phosphorylation has been lucidly discussed in another chapter of this book. If we accept the basic concept of the chemiosmotic hypothesis[22], it provides us with a specific view of the phenomenon of respiratory control. The translocation of protons and generation of a membrane potential that take place during respiration are self-limiting processes. In order for electron transport to continue, the proton gradient must be dissipated and the membrane potential collapsed.

A rather simple model system suitable for a classroom demonstration can be used to illustrate the point[23]. A crude commercial mixture of soybean phospholipids is suspended in 2% sodium cholate, sonicated in the presence of 0.1 M phenol red and dialysed against dilute buffer to remove cholate and external dye. When a small pulse of dilute HCl is added to the red vesicles, little or no change in colour can be seen indicating that (a) the dye is inside the vesicles and (b) phospholipid vesicles are impermeable to protons. When the membrane potential is collapsed with K^+ and valinomycin or when the proton gradient is dissipated, by a proton ionophore such as 1799, a very slow change of indicator colour can be observed. However, when valinomycin and the proton ionophore are added together, the vesicles turn yellow instantaneously. The interpretation of this experiment is given in Figure 5.1. In the presence of an uncoupler, proton movement into the vesicles is facilitated but restricted by the formation of a membrane potential. In the presence of K^+ and valinomycin the membrane potential is collapsed but proton movement is restricted by the low proton permeability of the phospholipid bilayer. When 1799 and valino mycin are added together, protons move in with the help of the uncoupler ionophore and the membrane potential is

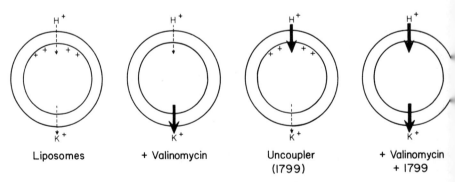

Figures 5.1   Mode of action of uncouplers and ionophores

collapsed by the outward movement of the charged complex of K^+–valinomycin.

A more sophisticated model for respiratory control can be prepared with cytochrome oxidase as translocator of electrons[24, 25]. This system is representative of the respiratory control observed in mitochondria since the orientation of the oxidase in the reconstituted vesicles is such that externally reduced cytochrome c is oxidised. Such vesicles are prepared by mixing cytochrome oxidase with phospholipids dissolved in 2% cholate and then removing excess cholate by dialysis. Alternatively, without cholate, phospholipids and cytochrome oxidase are briefly exposed to sonic oscillation, thus forming cytochrome oxidase vesicles with respiratory control[26]. These vesicles oxidise reduced cytochrome c at a very low rate and there is little or no stimulation on addition of valinomycin or nigericin alone. When both are added respiration is stimulated 5–12-fold.

The respiratory chain in mitochondria is asymmetrically assembled in such a manner that, during respiration, protons are translocated from inside to the outside (toward the outer mitochondrial membrane). According to the chemiosmotic hypothesis the proton motive force which is maintained by oxidation of substrate via the respiratory chain is used for ATP generation by the oligomycin-sensitive ATPase, which translocates protons back to the matrix side of the inner mitochondrial membrane. The problem of control of this system is therefore closely linked to the efficiency of coupling. Assuming that one ATP is formed for two H^+ translocated[22], we may ask how a tight coupling is achieved. Obviously, in damaged mitochondria or in diseased mitochondria[1] with impaired respiratory control the efficiency of coupling is greatly reduced. How is respiratory control lost? An increased leakage of protons through the membrane might be responsible. This could be caused either by chemical changes in the phospholipids or in the proteins. It was shown[25] that in reconstituted vesicles the compositions of both the phospholipids and the proteins have a pronounced influence on the respiratory control. Cardiolipin markedly increased respiratory control, while the presence of hydrophobic proteins greatly diminished it. In partly resolved submitochondrial particles, additions of either low concentrations of oligomycin[27] or of coupling factors are required for respiratory control[28] and proton translocation[29].

In intact mitochondria, respiratory control is measured by comparing the rate of respiration in state 3 and 4. According to the widely accepted nomenclature of Chance and Williams[30] during state 3 respiration, oxygen, substrate, P_i and ADP are present in excess. In state 4 the phosphate acceptor system is exhausted and respiration proceeds at a slow controlled rate. Bovine heart mitochondria which have been kept frozen for one or two days show no difference in respiration on addition of ADP —they lack respiratory control. We observed[31] that, after incubation of such mitochondria for 15 min at 30 °C, the rate of respiration is markedly accelerated by addition of ADP. Since mitochondria that have not been frozen but kept at 0 °C for the same length of time exhibit stimulation by ADP, we are dealing here with a repairable freezing damage. It would be of interest to investigate whether the damage caused by freezing of mitochondria involves primarily the lipids or whether there is also damage to the proteins. It might be worthwhile to explore whether the loosely coupled mitochondria of patients with characteristic muscular disease[1] or other diseased mitochondria[32] could also be repaired by similar manipulations.

5.2.2.2 Control of the coupling device

As mentioned earlier, according to the chemiosmotic hypothesis the coupling device responsible for the generation of ATP operates by the reverse action of an ATP-driven proton pump. It utilises the proton motive force created by the respiratory chains to translocate protons back to the other side of the membrane. That the oligomycin-sensitive ATPase can indeed function as an ATP-driven proton pump was shown experimentally in a reconstituted system[33, 33a]. Several proteins are required for this process. Among them is a hydrophobic protein which probably represents the transmembranous proton channel and several coupling factors, including the F_1–ATPase.

It is always puzzling to students that mitochondria which generate ATP contain an active ATPase. Actually, well-preserved mitochondria do not have much manifest ATPase because ATP hydrolysis is controlled by a specific protein inhibitor[34, 35]. This protein has a molecular weight of about 10 000 and interacts with the ATPase in such a manner that ATP hydrolysis is blocked while ATP generation remains unimpaired. On ageing or manipulation of the mitochondria the inhibitor dissociates from the enzyme and the ATPase becomes manifest. It is quite likely that dissociation of the inhibitor has physiological significance in view of the observation that it takes place during oxidation[36]. The effect of an increased mitochondrial ATPase activity on glycolysis will be discussed later.

5.2.2.3 Natural uncouplers and loose coupling

It has been known for many years that naturally occurring fatty acids are uncouplers of oxidative phosphorylation[37]. In terms of the chemiosmotic hypothesis they act as hydrophobic proton ionophores which collapse the proton gradient. The high degree of mobility of the unsaturated fatty acid may account for the observation that they are more effective uncouplers than saturated fatty acids[38]. There is remarkably little known about the possible role of free fatty acids in the control of energy metabolism. It appears that loss of respiratory control and uncoupling of oxidative phosphorylation by fatty acids is responsible for heat production by mitochondria in the brown fat of animals exposed to cold[39-41]. The involution of the mammary gland appears to be sparked by the appearance of free fatty acids that induce uncoupling[42].

Oxidative phosphorylation is also profoundly influenced by thyroid hormone. Removal of the thyroid gland or administration of thyroxine affects the rate of respiration of muscle mitochondria and thereby the net rate of ATP production. Thyroidectomy reduces oxygen uptake; thryroxine in either normal or thyroidectomised animals greatly accelerates respiration[43, 44]. A number of complex experimental observations such as the sensitivity of mitochondrial respiration to uncouplers, which falls and rises with the activity of thyroid hormone[45], might be interpretable in terms of changes in proton permeability.

Loosely coupled respiration has been operationally defined as respiration which is independent of the presence of ADP and phosphate yet still capable of generating ATP. A simple example are submitochondrial particles which respire in the absence of P_i and ADP yet catalyse oxidative phosphorylation with a P:O ratio approaching 3 provided an efficient phosphate acceptor system, e.g. glucose plus hexokinase, is present. Accepting the concept of proton impermeability as the key regulator of respiratory control, one can relate to it the efficiency of coupling. Various degrees of proton permeability can be visualised all the way from impermeability (tightly coupled respiration) to complete leakiness (uncoupled respiration). In between is loosely coupled respiration with various potentialities of phosphorylation dependent on the effectiveness with which the coupling device can cope with the proton leak.

The catecholamines which have a calorigenic action[45] have some claim as 'natural uncouplers', but their mode of action is complex and probably indirect via liberation of free fatty acids[41].

Perhaps a few words on 'pseudo-natural uncouplers' should be added, those administered by physicians. Some anaesthetics are uncouplers of oxidative phosphorylation[46]. Occasional fatalities during anaesthesia resulting in fulminating hyperthermia have the earmarks of an uncoupled death.

5.3 CONTROL OF ENERGY UTILISATION

We have defined earlier all reactions that result in generation of ADP and P_i from ATP as ATPase in the broadest sense. Included in this category are therefore all biosynthetic reactions that are ATP dependent and also all work that requires ATP. This embraces virtually all cellular metabolism as shown in Table 5.1. It is beyond the scope of this chapter to discuss control of biosynthetic processes, although they also utilise ATP. Moreover, particularly in slowly duplicating mammalian cells, these processes probably utilise only a small fraction of the energy that can be generated by the cell. Since, as emphasised earlier, the generation of energy is linked to its utilisation, we can conclude that the biosynthetic reactions do not contribute a significant share to the cellular ATPase activity. In line with this conclusion

Table 5.1   Potential sources of intracellular ADP and inorganic phosphate

Mitochondrial ATPase–F_1
Ca^{2+} ATPase of sarcoplasmic reticulum
Na^+K^+ ATPase of plasma membrane
Ca^{2+} ATPase of plasma membrane
Lysosomal ATPase
Viral ATPase
Biosynthetic processes
Muscular contraction and other work processes

is the general experience that enzyme activities of biosynthetic processes are usually described in terms of a specific activity in the picomole range while some work processes such as ion translocation have specific activities in the nanomole range.

In view of these considerations the following discussion is, with one exception, limited to the control of energy utilisation of the ATP by the work processes listed in Table 5.1.

5.3.1   Control of pump ATPases

There is remarkably little known about the control mechanisms of ATP-driven pumps, but we know that they are tightly coupled. For example, the Ca^{2+} pump of the sarcoplasmic reticulum hydrolyses one ATP for two Ca^{2+} translocated and we speak of a Ca^{2+}/ATP ratio of 2. We prefer the use of this ratio to the more frequently used reciprocal ratio of ATP/Ca^{2+} because the former becomes smaller with decreasing efficiency. Thus like the P/O ratio of oxidative phosphorylation the Ca^{2+}/ATP ratio of sarcoplasmic reticulum is an expression of the efficiency of operation. The Na^+K^+ pump of the plasma membrane also operates with a high efficiency and since the biochemistry of these two pumps is better understood than the proton pump of mitochondria or that of other pumps, we shall discuss them in greater detail.

The Ca^{2+} and Na^+K^+ pumps have certain features in common[47-50]. They both contain an ATPase that requires Mg^{2+}. The Ca^{2+} pump ATPase is activated by Ca^{2+}, the Na^+K^+ ATPase by Na^+ and K^+. The ATPases isolated in highly purified forms react with ATP to form phosphorylated proteins. In both cases it is an acyl phosphate in an aspartyl residue of a protein with a molecular weight of about 100 000[51,52]. Under the appropriate ionic conditions[48,56] both form the same acyl phosphate in the presence of inorganic phosphate without ADP or ATP being present. Both pumps can be operated in reverse utilising an ion gradient to generate ATP[53-55].

Thus a great deal is known about the molecular events that take place during the action of these pump enzymes, particularly when compared with the meagre information available about the operation of the proton pump of mitochondria. In contrast, we know much more about the control of the mitochondrial pump while virtually nothing is known about the control of the other ion pumps.

Like well-preserved mitochondria, freshly isolated sarcoplasmic reticulum vesicles have a high degree of efficiency with a Ca^{2+}/ATP ratio of 2. We have studied the mechanism of action and control of this pump by incorporating a highly purified preparation of Ca^{2+} ATPase[57] into liposomes in the presence of high concentrations of phosphate or oxalate[58]. Earlier preparations of these vesicles translocated Ca^{2+} in the presence of ATP at a rapid rate but with poor efficiency and considerable variability was encountered with different ATPase preparations. This induced us to study the efficiency of operation under a variety of conditions and methods of reconstitution[59]. Ca^{2+}/ATP ratios approaching 2 are now frequently obtained with reconstituted vesicles and it appears that a factor, perhaps a subunit of the ATPase similar to the mitochondrial ATPase inhibitor[54], plays a role in the control of the efficiency of Ca^{2+} translocation. That such a factor exists is very likely in view of the high turnover numbers of the purified ATPases from sarcoplasmic reticulum or from plasma membranes compared with the ATPase activity or rate of ion movement in the intact membrane or with membrane fragments. Such a calculation is permissible, particularly in the case of the sarcoplasmic reticulum ATPase which is widely believed to represent more than half of the total protein present in these organelles. Thus the purification achieved in preparing this enzyme[57] must be primarily concerned with the removal of minor constituents and the large increase in specific activity must be caused either by removal of an ATPase inhibitor or by conformational changes in the enzyme. Similar considerations apply to highly purified preparations of the Na^+K^+ ATPase. We shall return to the discussion of ion pump ATPases and their control[60, 61] in the section on tumor glycolysis.

5.3.2   Control of actomyosin ATPase

There is a vast literature on the ATPase activity associated with actomyosin. The complexity of its regulation[62, 63] accounts for some of the confusions of the past, but a reasonably clear picture is now emerging.

The key feature is the control of muscular contraction and of the actomyosin ATPase by Ca^{2+}. Ebashi and his collaborators[64] have shown that a complex of tropomyosin and troponin is required to confer Ca^{2+} sensitivity to actomyosin. Troponin contains three subunits required for its function[65]. One of them is a heat stable protein (Tp I) with a molecular weight of 24 000[66] which inhibits the hydrolysis of ATP by actomyosin ATPase (but not myosin ATPase). A second component of troponin has a molecular weight of 18 000 (Tp C) and is responsible for Ca^{2+} binding and conferral of Ca^{2+} sensitivity. The third component (mol. wt. 37 000) appears to enhance Ca^{2+} sensitivity and to link the two other components of troponin to tropomyosin[65] and perhaps also interacts with F-actin. The inhibitory component (Tp I) is counteracted by the Ca^{2+} sensitivity component (Tp C). A complex consisting of one tropomyosin and one troponin (three components) can regulate the activity of at least six actin monomers.

There is also regulation at the level of the myosin molecule itself. Rabbit skeletal myosin consists of two large polypeptides (mol. wt. 200 000) and several light chains (16 000–25 000) which are not covalently attached. Removal of the light chains results in loss of ATPase which can be reconstituted[67]. The role of some of the more readily dissociated light chains which can be removed without loss of ATPase activity remains still obscure[68]. In scallop myosin, removal of a particular light chain from myosin with EDTA results in a loss of Ca^{2+} sensitivity which can be restored by reconstitution[69]. This component inhibits the ATPase activity of scallop myosin activated by actin in the absence of calcium and therefore functions as a regulatory subunit. Similar desensitised preparations of actomyosin free of regulatory subunits have been prepared from rabbit skeletal muscle.

According to a recent report[70] the inhibitory protein of troponin inhibits also the ATPase activity of mitochondria. The kinetic properties of this inhibition are remarkably similar to those observed with the mitochondrial inhibitor, although comparatively much larger amounts of Tp I are needed to inhibit.

5.3.3   Control of ribosomal GTPase

The GTP hydrolysis catalysed by ribosomes in the presence of G factor is inhibited by a protein obtained from the wash of *E. coli* ribosomes[71]. The protein is heat stable, has a molecular weight of about 20 000 and is very sensitive to trypsin. The inhibitor is clearly a control protein since at low concentration it inhibits the GTPase activity without interfering with protein synthesis by the ribosomes[72].

The similarity between the various ATPase control proteins is rather striking. They are relatively small proteins (mol. wt. 10 000–24 000), heat stable, very sensitive to trypsin digestion and specifically interfere with the reactivity of the protein with water (ATP or GTP hydrolysis) without impairing its physiological transfer activity. Some of these control proteins may be so loosely bound that they are removed by simple washing (as in the case of ribosomal protein) or so tightly bound to the protein that they can be released from the protein only by heating in the presence of a detergent, as in the case of the chloroplast ATPase inhibitor[72a]. As mentioned earlier, the properties of the dissociable ATPase inhibitor of mitochondria is somewhere between these two extremes. The observation of the inhibition of mitochondrial ATPase by the troponin component Tp I re-emphasises the similarity between these control proteins and suggests that further cross-testing between the different inhibitors of ATPase activity might be profitable.

In spite of their complexity one can clearly discern similarities between the energy transducing mechanism involved in the generation of ATP in mitochondria and in the ATP utilisation during muscular contraction. The ATPase of F_1 and of myosin have properties in common that we have pointed out[73] when we first described F_1. The presence of two major subunits and several small subunits, some of them with regulatory function, are additional common features discovered since. But more striking are the similarities between the functional complex actomyosin–tropomyosin–troponin in muscular contraction and the oligomycin-sensitive ATPase with its multiple coupling factors in oxidative phosphorylation. The observation that one of the regulatory subunits can actually participate in both systems[70] suggests that some of the similarities are not accidental.

5.4   CONTROL OF ENERGY PRODUCTION AND UTILISATION IN TUMOR CELLS

In discussing this subject we have departed from the broader coverage of the earlier sections in this chapter and have included details of an experimental approach. The reason for this departure is the fact, mentioned earlier, that particularly the control of energy utilisation is a rather neglected area of research, yet of great physiological importance. Since our own efforts have concentrated on controls of energy production and utilisation in tumor cells, this will be the major subject of discussion.

5.4.1   The Warburg effect

Fifty years ago Warburg reported that tumors have a high aerobic glycolysis[74]. This simple observation was soon complicated on the one hand by the introduction of complex calculations and quotients which confused the issue, and on the other hand by Warburg's insistence that the phenomenon is caused by a defect in the respiratory mechanism of the cancer cell. This generalisation was clearly erroneous and the polemics which followed had the curious effect that the phenomenon of acid production in tumor cells has been largely neglected. The 'conquest of cancer' programme of the United States government has not devoted even a small segment of its complex wheel chart to this problem.

The history of many important discoveries is said to take place in three stages. First, people say it's not true, then they say it's true but unimportant, finally they say it's true and important, but not new. Warburg's discovery was stuck in the second stage for so long that it became not only worn out and old, but generally considered to be unimportant.

There is a reasonably good correlation between the rate of tumor growth and aerobic glycolysis[75]. Yet there can be little doubt[76] that certain normal tissues, as well as cells grown in tissue culture, have a high aerobic glycolysis. Obviously, lactic acid production alone is not sufficient to make a cell malignant if it does not have the required potentials for rapid growth and infiltration. But the key questions that should be raised and are pertinent to this chapter: Do tumor cells have a defect in their energy transducing systems? Is this defect reponsible for the high lactic acid production? Is the lactic acid production *per se* a feature contributory to malignancy or simply an unimportant secondary or tertiary consequence of a primary defect in replication?

The reviewer has proposed as a working hypothesis[77] that (a) there can be different lesions that can induce a high aerobic glycolysis; (b) that the resulting low intracellular pH and other consequences of the same lesion have a profound effect on the control of cellular metabolism and growth; and (c) that the excretion of lactic acid by the tumor may damage the host cells and facilitate infiltration.

A systematic analysis was initiated to determine the lesion that is responsible for the high aerobic glycolysis in various tumor cells. The second step was to find a way to repair the lesions and finally to establish whether a repair of the lesion *in vivo* has any influence on the malignancy of tumor growth.

5.4.2   ATPases and tumor glycolysis

We have shown many years ago[10, 78] that in tumor cells that we have analysed the glycolytic enzymes are present in excess and that the generation of inorganic phosphate and ADP are rate-limiting factors in glucose utilisation. Which of the various ATPases listed in Table 5.1 contribute to the supply of ADP and phosphate? The first tumor cell we selected for study was the Ehrlich ascites tumor cell. Since it was necessary to analyse the intact cells, we had to resort to the use of inhibitors. Some inhibitors have sufficient specificity to permit their use, provided mutliple approaches and precautions are taken, as will be illustrated in the analysis of the ascites tumors. Ruta-mycin, which inhibits mitochondrial ATPase, had no effect on the glycolysis of the ascites cells. This negative finding was ambiguous for two reasons. One is the possibility that rutamycin may not permeate to the mitochondria; the second is that the effect of rutamycin may be complex. By inhibiting mito-chondrial ATPase activity it should reduce the rate of glycolysis, but by eliminating oxidative phosphorylation it should enhance glycolysis. More-over, high concentrations of rutamycin also inhibit the Na^+K^+ ATPase of the plasma membrane. The question of permeability was readily settled by showing that in the presence of dinitrophenol, which stimulated glycolysis, rutamycin was an effective inhibitor of lactate production[60]. The dual effect of rutamycin was more difficult to evaluate. The more decisive finding was a marked inhibition of glycolysis by ouabain. This compound is generally considered to be a rather specific inhibitor of the Na^+K^+ ATPase[48,79]. Moreover, an analysis of Rb^+ uptake by the ascites tumor revealed a close agreement between the inhibition of transport and of glycolysis[60]. To rule out the possibility proposed by some investigators[80] that the inhibition of glycolysis by ouabain was caused by a lack of potassium, we showed that dinitrophenol-stimulated glycolysis was not sensitive to ouabain under the same experimental conditions.

A variety of cell lines and tumors were tested in the presence of rutamycin, 2,4-dinitrophenol and ouabain[61]. As can be seen from Table 5.2, some very distinct patterns emerge. Some cells like the Ehrlich ascites tumors and the neuroblastoma cells are very sensitive to ouabain, indicating that the Na^+K^+ ATPase supplies the ADP and P_i required for glycolysis. None of the other four cell lines were inhibited by ouabain. Rutamycin inhibited the glycolysis in 3T3 and polyoma virus transformed 3T3 fibroblasts, suggesting that in these cells the mitochondrial ATPase sustains glycolysis. The observation that glycolysis of neuroblastoma cells is inhibited by rutamycin as well as by ouabain is difficult to evaluate, since the Na^+K^+ ATPase of some cells is inhibited at the concentration of rutamycin used in these experiments. More specific inhibitors of mitochondrial ATPase are therefore needed. Atractyl-oside[81] and bongkrekic acid[82], which are both specific inhibitors of mitochon-drial nucleotide transport, could qualify but the former is deficient in permeability properties and the latter in general availability.

As might be expected, the stimulation of glycolysis by 2,4-dinitrophenol was most pronounced in cells which did not have an activated mitochondrial ATPase. However, the mode of action of dinitrophenol is also complex since it not only inhibits oxidative phosphorylation but also stimulates the ATPase activity of mitochondria. The latter component seems to be the more important one since rutamycin which inhibits oxidative phosphoryla-tion stimulated glycolysis only slightly or not at all, while dinitrophenol stimulated both anaerobic as well as aerobic glycolysis of ascites tumor cells.

177

Table 5.2 Effect of inhibitors on lactate formation in various cell lines

| Additions | Ascites cells | Lactate formation/µmol 30 min^{-1} mg^{-1} protein | | | | |
		3T3	PY–3T3	Neuroblastoma	BHK	PY–BHK
None	0.35	0.41	0.86	0.98	0.6	0.7
Ouabain (1 mM)	0.12	0.60	0.70	0.57	0.5	0.7
Dinitrophenol (0.1 mM)	1.28	0.69	1.40	1.63	1.5	2.0
Rutamycin (8 µg ml^{-1})	0.45	0.18	0.27	0.60	1.2	1.7

It is therefore curious that *in vitro* studies with mitochondria isolated from tumors failed to reveal a dinitrophenol-activated ATPase[83]. These interesting observations suggests that a change has taken place in the energy transducing system in the course of the isolation of the mitochondria.

Ouabain had little or no effect on the glycolysis of baby hamster kidney cells and polyoma virus transformed cells; rutamycin actually stimulated glycolysis. These findings suggest that there is either another ATPase sustaining glycolysis or that the Na^+K^+ pump is insensitive to ouabain. The latter possibility seems to be ruled out by the finding that Rb^+ uptake was inhibited by ouabain. Quercetin, a bioflavonoid which inhibits all membranous ATPase thus far tested, was an effective inhibitor of glycolysis in baby hamster kidney cells. We shall discuss the mode of action of flavonoids later.

5.4.3   Repair of the lesion

Ouabain and rutamycin are not suitable compounds to test the role of glycolysis in the growth of tumors. They inhibit essential energy transducing processes in normal cells and are therefore quite toxic. To test the hypothesis[77] that the acidity or other intracellular changes induced by the high aerobic acid formation contribute to the malignant properties of tumors, we need compounds which are not toxic for normal cells but specifically repair the lesion in tumor cells responsible for the high ATPase activity. We propose that the high ATPase activity that sustains high aerobic glycolysis in tumor cells is caused by a membranous lesion that renders the pump less efficient and lowers the ion/ATP ratio. We therefore consider the possibilities that either the ATPase itself or the ATPase inhibitor proteins discussed earlier may be affected. Either less inhibitor is available or it is altered so that it becomes a less effective control mechanism for the pump.

The natural ATPase inhibitors would be the obvious material to test the validity of this hypothesis. However, attempts to inhibit mitochondrial ATPase of intact tumor cells by mitochondrial ATPase inhibitor were not successful, presumably because this polypeptide (molecular weight of 10 000) does not pass the permeability barrier of the membranes that block the access to the mitochondrial ATPase.

We therefore searched for small molecular, hydrophobic compounds that inhibit ATPase activity without inhibiting the energy transducing processes. Such compounds were found among the group of bioflavonoids[61, 84] shown in Table 5.3. Several representatives of this group inhibit mitochondrial ATPase and, at low concentrations, have no effect on oxidative phosphorylation. We have explored the effect of a large number of flavonoids and related compounds on the glycolysis of tumor cells as well as on ATPase activities and we are slowly beginning to learn something about the relationship between the structure and function of these compounds. Blocking of the hydroxyl group at the three positions by a glucoside completely eliminated the effectiveness of quercetin as an inhibitor of glycolysis or ATPase activity. Introduction of a charged sulphonate group at several other positions also results in inactivation. Dihydroxyflavone and trihydroxyflavone were inactive, but a tetrahydroxychalcone did show activity. It appears from these studies that an OH group near the carbonyl oxygen is required and that the compound must have hydrophobic properties. The former property correlates with the chelating capacity of these compounds. The latter may be required to interact with the active site of the ATPase. The possibility of a metal firmly bound at the active site of the various ATPases is now under investigation.

Table 5.3   Effect of flavones on mitochondrial and plasma membrane ATPases

Name	Hydroxyl groups at	Concentration	Mitochondrial	Na^+K^+ ATPase
Quercetin	3,3′,4′,5,7	$8\ \mu g\ ml^{-1}$	75%	85%
Myricetin	3,3′,4′,5,6,7	$8\ \mu g\ ml^{-1}$	73%	86%
Luteolin	3′,4′,5,7	$8\ \mu g\ ml^{-1}$	21%	66%
Fisetin	3,3′,4′,7	$16\ \mu g\ ml^{-1}$	43%	60%
Galangin	3,5,7	$40\ \mu g\ ml^{-1}$	0	48%
Chrysin	5,7	$40\ \mu g\ ml^{-1}$	0	0
Rutin	Quercetin-3-glycoside	$40\ \mu g\ ml^{-1}$	0	0

The inhibition of aerobic glycolysis of 3T3 fibroblasts and of polyoma virus transformed 3T3 cells by flavonoids was expected since these compounds inhibit mitochondrial ATPase. Surprisingly, glycolysis in ascites tumor and other cells which utilised ADP and phosphate generated by other pumps were also found to be sensitive to quercetin. Since previous work[60] has clearly implicated the Na^+K^+ ATPase as source of P_i and ADP, e.g. in Ehrlich ascites tumor cells, this finding suggested that the plasma enzyme is also susceptible to low concentrations of flavonoids. Highly purified preparations of the ATPase from dog and sheep kidney, or from the electric eel, were very sensitive to quercetin. Moreover, the concentration of quercetin that inhibited glycolysis or the ATPase had no effect on the uptake of rubidium by the intact cells. This mode of action is quite different from that of ouabain, which inhibits both glycolysis and Rb^+ uptake[60]. These observations suggested that quercetin controls the ATPase of the Na^+K^+ pump without interfering with the energy transducing process. In other words, its action on the Na^+K^+ pump is similar to that of the natural mitochondrial inhibitor of the ATPase which participates in oxidative phosphorylation. It should be mentioned that quercetin had no effect on the glycolysis of an ascites tumor extract which was supplied with an excess of ADP and phosphate.

Since the major source of ADP in Ehrlich ascites tumor cells is derived from the ATPase activity of the plasma membrane, and since one ATP must be cleaved for each lactate that is produced from glucose, one can estimate the efficiency of the Na^+K^+ pump operation from the rate of Rb^+ uptake and lactate formation measured under identical experimental conditions. The ratio of Rb^+/lactate should correspond to the ratio of Rb^+/ATP and be indicative of the efficiency of the pump. In Ehrlich ascites tumor cells the observed Rb^+/ATP ratio[60] was below 1, indicated a low efficiency. We observed recently[85] that quercetin has a pronounced effect on the ratio by inhibiting the formation of lactate without affecting the transport of Rb^+. In fact, sometimes the ratio exceeded the expected value of 2, suggesting that at the low rate of glycolysis another process effectively competed for the available ADP and P_i. Since oxidative phosphorylation was an obvious candidate for such a competition, the assays were repeated in the presence of dinitrophenol to eliminate the mitochondrial utilisation of ADP. The bioflavonoids which inhibit the mitochondrial as well as the plasma membrane ATPase raised the Rb^+/lactate ratio under these conditions to a value close to 2. It can therefore be concluded that quercetin permits a more efficient operation of the Na^+K^+ pump and may be looked upon as 'coupler' of a loosely coupled Na^+K^+ pump.

An important drawback of quercetin became apparent when it was found that calf serum, which is a constituent of the growth medium, eliminated the inhibitory effect on glycolysis. Two components in the serum were found to prevent the action of quercetin on the glycolysis of ascites tumor cells. One was serum albumin, the second sodium bicarbonate. Bovine serum albumin also prevented and reversed the effect of quercetin on soluble ATPase (e.g. F_1), while bicarbonate only prevented the inhibition in intact cells. However, experiments on the effect of quercetin on cells grown in tissue culture could still be carried out by avoiding bicarbonate and by reducing the content of serum in the growth medium to a minimum compatible with growth. Under these conditions quercetin inhibited markedly the growth of several cells tested, including 3T3, polyoma transformed 3T3 and two strains of leukemia cells (L1210 and P 388). The latter two are among the cancer cells used in the chemotherapy survey of the National Cancer Institute, where currently quercetin is being tested *in vivo*.

We have included these details on the effect of bioflavonoids on the energy metabolism of tumor cells only to illustrate the type of approach that can be used to analyse the complex mechanisms that control cellular energy transducing processes. It is also apparent that we need to have much more fundamental information on the operation of the pumps. Reconstituted systems of pumps may be of considerable help in elucidating the mechanisms of the pump proper as well as of the controls that permit the efficient operation of the pumps. Eventually the components of the pumps themselves and of their control systems will have to be isolated in pure form and characterised by physical and chemical methods. Let us face the fact that when it comes to the specific question, how ions are transported via the ATPases from one side of the membrane to the other, we not only have no answer, but only limited approaches to obtain an answer. The field is wide open for a new generation of investigators.

Acknowledgement

The work described in this review was supported by United States Public Health Service Grants CA-08964 and CA-14454 from the National Cancer Institute and by the American Cancer Society Grant BC-156.

References

1. Luft, R., Ikkos, D., Palmieri, G., Ernster, L. and Afzelius, B. (1962). *J. Clin. Invest.*, **41**, 1776
2. Worsfold, M., Peter, J. B. and Dunn, R. F. (1969). *Proc. Int. Congr., Milan*, 303
3. Sidbury, J. B. Jr., Smith, E. K. and Harlan, W. (1967). *J. Pediat.*, **70**, 8
4. Dodds, E. C. and Greville, G. D. (1934). *Lancet I*, 398
5. Loomis, W. F. and Lipmann, F. (1948). *J. Biol. Chem.*, **173**, 807
6. Krimsky, I. and Racker, E. (1963). *Biochemistry*, **2**, 512
7. Lipmann, F. (1946). In *Advances in Enzymology*, Vol. VI, 231 (F. F. Nord, editor) (New York: Interscience)
8. Allison, W. S. and Benitez, L. V. (1972). *Proc. Nat. Acad. Sci. USA*, **69**, 3004
9. Racker, E. and Krimsky, I. (1952). *J. Biol. Chem.*, **198**, 731
10. Wu, R. and Racker, E. (1959). *J. Biol. Chem.*, **234**, 1036
11. Lardy, H. A. and Parks, R. E. Jr. (1956). In *Enzymes: Units of Biological Structure and Function*, 584 (O. H. Gaebler, editor) (New York: Academic Press)
12. Passoneau, J. B. and Lowry, O. H. (1962). *Biochem. Biophys. Res. Commun.*, **7**, 10
13. Rose, I. A., Warms, J. V. B. and O'Connel, E. L. (1964). *Biochem. Biophys. Res. Commun.*, **15**, 33
14. Uyeda, K. and Racker, E. (1965). *J. Biol. Chem.*, **240**, 4682
15. Lynen, F., Hartmann, G., Netter, K. F. and Schuegraf, A. (1959). In *Ciba Foundation Symposium on the Regulation of Cell Metabolism*, 256 (G. E. W. Wolstenholme and C. M. O'Conner, editors) (London: Churchill)
16. Williams, J. R., Herczeg, B. A., Coles, H. S. and Cheung, W. Y. (1967). *J. Biol. Chem.*, **242**, 5119
17. Uyeda, K. and Racker, E. (1965). *J. Biol. Chem.*, **240**, 4689
18. Lynen, F. and Königsberger, R. (1951). *Justus Liebigs Ann. Chem.*, **573**, 60
19. Rapoport, S. M. (1974). Personal communication
20. Tornheim, K. and Lowenstein, J. M. (1973). *J. Biol. Chem.*, **248**, 2670
21. Racker, E. (1954). In *Advances in Enzymology*, Vol. XV, 141 (F. F. Nord, editor) (New York: Interscience)
22. Mitchell, P. (1966). *Biol. Rev. Cambridge Phil. Soc.*, **41**, 445
23. Racker, E. (1974). In *Dynamics of Energy-transducing Membranes*, 269 (Ernster, Estabrook and Slater, editors) (Amsterdam: Elsevier)
24. Hinkle, P. C., Kim, J.-J. and Racker, E. (1972). *J. Biol. Chem.*, **247**, 1338
25. Racker, E. (1972). *J. Membrane Biol.*, **10**, 221
26. Racker, E. (1973). *Biochem. Biophys. Res. Commun.*, **55**, 225
27. Lee, C. P. and Ernster, L. (1968). *Europ. J. Biochem.*, **3**, 391
28. Cockrell, R. S. and Racker, E. (1969). *Biochem. Biophys. Res. Commun.*, **35**, 414
29. Hinkle, P. and Horstman, L. L. (1971). *J. Biol. Chem.*, **246**, 6024
30. Chance, B. and Williams, G. R. (1956). *Adv. Enzymol.*, **17**, 65
31. Racker, E. and Horstman, L. L. (1972). In *Energy Metabolism and the Regulation of Metabolic Processes in Mitochondria*, 1 (M. A. Mehlman, R. W. Hanson, editors) (New York: Academic Press)
32. Peter, J. B., Stempel, K. and Armstrong, J. (1969). *Proc. Int. Congr., Milan*, 228
33. Kagawa, Y. and Racker, E. (1971). *J. Biol. Chem.*, **246**, 5477
33a. Kagawa, Y., Kandrach, A. and Racker, E. (1973). *J. Biol. Chem.*, **248**, 676
34. Pullman, M. E. and Monroy, G. C. (1963). *J. Biol. Chem.*, **238**, 3762
35. Racker, E. and Horstman, L. L. (1967). *J. Biol. Chem.*, **242**, 2547
36. Van de Stadt, R. J., De Boer, B. L. and Van Dam, K. (1973). *Biochim. Biophys. Acta*, **292**, 338

37. Pressman, B. C. and Lardy, H. A. (1956). *Biochim. Biophys. Acta*, **21,** 458
38. Borst, P., Loos, J. A., Christ, E. J. and Slater, E. C. (1962). *Biochim. Biophys. Acta*, **62,** 509
39. Guillory, R. J. and Racker, E. (1968). *Biochim. Biophys. Acta*, **153,** 490
40. Rafael, V. J., Klaas, D. and Hohorst, H.-J. (1968). *Z. Physiol. Chem.*, **349,** 1711
41. Prusiner, S. B., Cannon, B. and Lindberg, O. (1968). *Europ. J. Biochem.*, **6,** 15
42. Nelson, W. L., Butow, R. A. and Ciaccio, E. I. (1962). *Arch. Biochem. Biophys.*, **96,** 500
43. Ernster, L. (1965). *Fed. Proc.*, **24,** 1222
44. Hoch, F. L. (1971). *Energy Transformations in Mammals: Regulatory Mechanisms, Physiological Chemistry*, (London: W. B. Saunders)
45. Griffith, F. R. Jr. (1951). *Physiol. Rev.*, **31,** 151
46. Snodgrass, P. J. and Piras, M. M. (1966). *Biochemistry*, **5,** 1140
47. Post, R. L., Sen, A. K. and Rosenthal, A. S. (1965). *J. Biol. Chem.*, **240,** 1437
48. Schwartz, A., Lindenmayer, G. E. and Allen, J. C. (1972). In *Current Topics in Membranes and Transport*, Vol. 3, 1 (New York: Academic Press)
49. Martonosi, A. (1973). In *Current Topics in Membranes and Transport*, Vol. 3, 83 (New York: Academic Press)
50. Hasselbach, W. (1972). In *Molecular Bioenergetics and Macromolecular Biochemistry*, 149 (H. H. Weber, editor) (Berlin: Springer-Verlag)
51. Post, R. L. and Küme, S. (1973). *J. Biol. Chem.*, **248,** 6993
52. Bastide, F., Meissner, G., Fleischer, S. and Post, R. L. (1973). *J. Biol. Chem.*, **248,** 8385
53. Glynn, I. M. and Lew, V. L. (1969). In *Membrane Proteins* (*New York Heart Association Symposium*) 289 (Boston: Little, Brown)
54. Makinose, M. and Hasselbach, W. (1971). *FEBS Lett.*, **12,** 271
55. Panet, R. and Selinger, Z. (1972). *Biochim. Bioph. Acta*, **255,** 34
56. Masuda, H. and de Meis, L. (1973). *Biochemistry*, **12,** 4581
57. MacLennan, D. H. (1970). *J. Biol. Chem.*, **245,** 4508
58. Racker, E. (1972). *J. Biol. Chem.*, **247,** 8198
59. Racker, E. and Eytan, E. (1973). *Biochem. Biophys. Res. Commun.*, **55,** 174
60. Scholnick, P., Lang, D. and Racker, E. (1973). *J. Biol. Chem.*, **248,** 5175
61. Suolinna, E., Lang, D. and Racker, E. (1974). *J. Nat. Cancer Inst.*, **53,** 1515
62. Huxley, H. E. (1972). *Cold Spring Harbor Symposia on Quantitative Biology*, Vol. 37, 689 (New York: Cold Spring Harbor)
63. Tonomura, Y. (1972). *Muscle Proteins, Muscle Contraction and Cation Transport* (Tokyo: University of Tokyo Press)
64. Ebashi, S. and Endo, M. (1968). *Progr. Biophys. Mol. Biol.*, **18,** 123
65. Greaser, M. L. and Gergely, J. (1973). *J. Biol. Chem.*, **248,** 2125
66. Schaub, M. C. and Perry, S. V. (1971). *Biochem .J.*, **123,** 367
67. Dreizen, P. and Richards, D. H. (1972). *Cold Spring Harbor Symposia on Quantitative Biology*, Vol. 37, 29 (New York: Cold Spring Harbor)
68. Gazith, J., Himmelfarb, S. and Harrington, W. F. (1970). *J. Biol. Chem.*, **245,** 15
69. Kendrick-Jones, J., Szentkiralyi, E. M. and Szent-Györgyi, A. G. (1972). *Cold Spring Harbor Symposia on Quantitative Biology*, Vol. 37, 47 (New York: Cold Spring Harbor)
70. Tamaura, Y., Yamazaki, S., Hirose, S. and Inada, Y. (1973). *Biochem. Biophys. Res. Commun.*, **53,** 673
71. Kuriki, Y. and Kanno, K. (1972). *Biochim. Biophys. Res. Commun.*, **48,** 700
72. Kuriki, Y. and Yoshimura, F. (1974). *J. Biol. Chem.*, **249,** 7166
72a. Nelson, N., Nelson, H. and Racker, E. (1972). *J. Biol. Chem.*, **247,** 7657
73. Pullman, M. E., Penefsky, H. S. and Datta, A. (1960). *J. Biol. Chem.*, **235,** 3322
74. Warburg, O. (1926). *Uber den Stoffwechsel der Tumoren* (Berlin: Springer-Verlag)
75. Wenner, C. E. (1967). In *Advances in Enzymology*, Vol. 29, 321 (F. F. Nord, editor) (New York: Interscience)
76. Krebs, H. A. (1970). In *Essays in Biochemistry*, Vol. 7, p. 1 (P. N. Campbell and F. Dickens, editors) (New York: Academic Press)
77. Racker, E. (1972). *American Scientist*, **60,** 56
78. Alpers, J. B., Wu, R. and Racker, E. (1963). *J. Biol. Chem.*, **238,** 2274
79. Schatzmann, H. J. (1953). *Helv. Physiol. Pharmacol. Acta*, **11,** 346
80. Poole, D. T., Butler, T. C. and Williams, M. E. (1971). *J. Membrane Biol.*, **5,** 261
81. Burni, A., Contessa, A. R. and Luciani, S. (1962). *Biochim. Biophys. Acta*, **60,** 301
82. Henderson, P. J. F. and Lardy, H. A. (1970). *J. Biol. Chem.*, **245,** 1319

83. Pedersen, P. L., Eska, T., Morris, H. P. and Catterall, W. A. (1971). *Proc. Nat. Acad. Sci. USA*, **68,** 1079
84. Lang, D. and Racker, E. (1974). *Biochim. Biophys. Acta*, **333,** 180
85. Suolinna, E., Buchsbaum, R. and Racker, E. (1975). *Cancer Research*, in the press

Index